# The Logica Yearbook
# 2012

# The Logica Yearbook 2012

Edited by

## Vít Punčochář
and
## Petr Švarný

© Individual author and College Publications 2013
All rights reserved.

ISBN 978-1-84890-110-0

College Publications
Scientific Director: Dov Gabbay
Managing Director: Jane Spurr

www.collegepublications.co.uk

Original cover design by Laraine Welch
Printed by Lightning Source, Milton Keynes, UK

---

All rights reserved. No part of this publication may be reproduced, stored in a retrieval system or transmitted in any form, or by any means, electronic, mechanical, photocopying, recording or otherwise without prior permission, in writing, from the publisher.

# Preface

Since 1987 when the Institute of Philosophy of Czechoslovak Academy of Sciences organized the symposium Logica '87 the Logica symposia have become one of the most traditional annual conferences devoted to logic. Since 1997 speakers in the symposia have had the opportunity of publishing their contributions in The Logica Yearbook series. Last year was no exception and so we have the pleasure of introducing the latest volume of the proceedings which contains some of the papers presented at Logica 2012.

Logica 2012, held at Hejnice Monastery (North Bohemia) from 18 to 22 June 2012, was organized by the Department of Logic in the Institute of Philosophy of the Academy of Sciences of the Czech Republic. As every year, the symposium brought together logicians from many countries and besides the invited talks—invited speakers were Sergei Artemov, Warren Goldfarb, David Makinson, and Barbara Partee—about thirty other papers devoted to the various branches of logic were presented.

Both the Logica symposium and The Logica Yearbook are the result of the joint effort of many people who deserve our warmest thanks. We are very grateful to the Institute of Philosophy and especially its director, Pavel Baran, for all their support. We would like to thank College Publications and its managing director Jane Spurr. Special thanks go to Petra Ivaničová who provided invaluable assistance to the organizers of the conference.

We are also very grateful to the staff of Hejnice Monastery and to Bernard Family Brewery of Humpolec which has traditionally sponsored the social programme of the symposium. Neither the publication of this volume, nor the conference Logica 2012 itself would be possible without the Grant Agency of the Czech Republic which provided significant support by financing the grant project no. P401/10/1279.

Last, but not least, we would like to thank all the authors for their exemplary collaboration during the editorial process.

Prague, May 2013

Vít Punčochář and Petr Švarný

# Contents

*Roberto Ciuni and Carlo Proietti:*
Probabilistic Semantics for a Discussive
Temporal Logic 1

*Christopher N. Foster:*
Overview of a Tarskian Solution to the Iterated
Prisoner's Dilemma 15

*Chris Fox:*
Axiomatising Questions 23

*Mark Jago:*
Are Impossible Worlds Trivial? 35

*Bjørn Jespersen:*
Alleged(ly) 51

*Neil Kennedy:*
On Modal Facts in Possible Worlds 65

*Paweł Łupkowski:*
Cooperative Question-responses and Question
Dependency 79

*David Makinson:*
Advice to the Relevantist Policeman 91

*Johannes Marti:*
  **Semantic Facts on Kripke Frames**   101

*Barbara H. Partee:*
  **The Starring Role of Quantifiers in the History of Formal Semantics**   113

*Victor Rodych:*
  **Wittgenstein Completeness**   137

*Igor Sedlár:*
  **From Pair Points to Pairs of Models**   159

*Hartley Slater:*
  **Towards Being**   171

*Petr Švarný:*
  **Flow of Time & Branching**   181

*Paul D. Thorn:*
  **Cognitivist Probabilism**   201

*Luca Tranchini:*
  **An Analogy in Dummett's Views on Truth-conditional and Proof-conditional Meaning Theories**   215

# Probabilistic Semantics for a Discussive Temporal Logic

ROBERTO CIUNI[1] AND CARLO PROIETTI[2]

**Abstract:** The paper introduces a probabilistic semantics for the paraconsistent temporal logic $Ab$ presented by the authors in a previous work on future contingents. Probabilistic concepts help frame two possible interpretations of the logic in question—a 'subjective' and an 'objective' one—and explaining the *rationale* behind both of them. We also sketch a proof-method for $Ab$ and address some considerations regarding the conceptual appeal of our proposal and its possible future developments.

**Keywords:** probability, discussive logic, future contingents

## 1 Introduction

In (Ciuni & Proietti, to appear) we introduce the paraconsistent temporal logic $Ab$. Our main motivation there is to explore a rather unusual theoretical option for future contingent propositions—being *both true and false*—and to investigate how this fares with the problem of *logical determinism*. $Ab$ has a strong affinity with Jaskowski's system J (see Jaśkowski, 1969) and its notion of *truth*, or *subtruth*, equates with "truth in at least one possible course of events", i.e. a dual of Thomason's supervaluationist notion of truth as "truth in every possible course of events" (see Thomason, 1970). An interesting virtue of our approach is that it can adequately express cases of 'retrogradation of truth'[3] and yet guarantee—for certain inference modes—a higher adherence to classical logic than supervaluationism.[4]

---

[1] This paper was written while the first author was a Humboldt Postdoctoral Fellow with the project 'A Tempo-Modal Logic for Responsibility Attribution in Many-Step Actions' (2011-2013) and while he was a collaborator of the project 'Logiche tempo-modali per gli agenti deontici' (University of Padova, Departments of Philosophy and Mathematics, 2011-2013).

[2] This paper was written while the second author was employed as a full-time researcher at the Department of Philosophy of Lund University as a memeber of the 'Knowledge and Information Quality' research group.

[3] 'Retrogradation of truth' is the logical process of evaluating at moment $m$ a formula which has been asserted at an *earlier* moment $m'$. For instance, we face a case of retrogradation of truth when we evaluate today the sentence "It was the case yesterday that there was going to be a sun eclipse in a day".

[4] See (Ciuni & Proietti, to appear), Theorem 5 and Theorem 6.

While our proposal had a technical focus, its conceptual underpinnings remained largely unexplored. In particular, this lack opens to an ambiguity between *two possible different readings* of the notion of subtruth, a *subjective* reading and an *objective* one, which in turn blurs the intuitive grasp of the semantics and—consequently—its potential for applications. These defects, however, can be emended.

In this paper we proceed as follows. Section 2 briefly introduces the semantics presented in (Ciuni & Proietti, to appear), its main formal features and sketches a proof method for it. Section 3 clarifies the ambiguity between the *objective* and the *subjective* notion of truth by detailing both options. Furthermore, we provide an alternative probabilistic semantics which gives a more natural insight to both readings (the reader can find the reason for this at the end of section 6 of (Ciuni & Proietti, to appear)).[5] We then prove that the notions of logical consequence defined for the two semantics are *extensionally equivalent* w.r.t. the class of countable structures. After this, we analyze closer the (probabilistic) subjective notion of truth—as *average expectation*—in terms of individual assignments of probability and show its affinity with the spirit of discussive logics. Finally, section 4 summarizes the results of the paper, discusses the main conceptual points, the virtues of the present proposal, and presents some conclusions.

## 2 A Discussive Logic for Future Contingents

The logic $Ab$ introduced in (Ciuni & Proietti, to appear) is based on a language $\mathcal{L}$ recursively defined over a set $\mathfrak{Prop}$ of atomic formulas $p, q, \ldots$ (intuitively representing immediate present tense sentences) using boolean ($\neg, \wedge, \vee, \rightarrow$) and temporal operators $F$ ("it will be the case that") and $P$ ("it has been the case that"). Following the standard, formulas of $Ab$ are evaluated on *trees* $\mathcal{T} := (T, <)$, where $T$ is a nonempty set of moments $m, m', m'', \ldots$ and $<$ is a strict ordering relation[6] which is *backward linear* and allows *forward branching*. A *history* $h$ is a maximal $<$-chain in $T$; $H$ is the set of histories in $\mathcal{T}$ and for every moment $m$, $H_m$ will designate the set of all histories 'passing through $m$' (that is s.t. $m \in h$).

---

[5] Probabilistic semantics for branching time have already been proposed in the field of model checking. A prominent example is (Baier & Kwiatkowska, 1998). The main difference is that the probability distribution there applies to *states*—which we call *moments* here—while it applies to *histories*—maximal chains of moments—here.

[6] That is, the relation is irreflexive, transitive and asymmetrical.

A preliminary move for building the subvaluationist semantics of $Ab$ is to define a "classical" *Ockhamist model* $\mathcal{M}$ as a pair $(\mathcal{T}, V)$, where $V$ is an evaluation function assigning to every propositional variable a subset of $T \times H$. Satisfaction is defined by the following clauses (see also Hasle & Øhrstrom, 1995)

$\mathcal{M}, (m, h) \models_{Ock} p \Leftrightarrow (m, h) \in V(p)$

$\mathcal{M}, (m, h) \models_{Ock} \neg \phi \Leftrightarrow \mathcal{M}, (m, h) \not\models_{Ock} \phi$

$\mathcal{M}, (m, h) \models_{Ock} \phi \wedge \psi \Leftrightarrow \mathcal{M}, (m, h) \models_{Ock} \phi$ and $\mathcal{M}, (m, h) \models_{Ock} \psi$

$\mathcal{M}, (m, h) \models_{Ock} P\phi \Leftrightarrow \exists m' < m$ such that $\mathcal{M}, (m', h) \models_{Ock} \phi$

$\mathcal{M}, (m, h) \models_{Ock} F\phi \Leftrightarrow \exists m' > m$ such that $\mathcal{M}, (m', h) \models_{Ock} \phi$

Also, in (Ciuni & Proietti, to appear) we impose the standard condition that atoms should receive the same evaluation relative to any history at a given moment. There may be reasons to relax this condition, allowing atoms to get different evaluations relative to different histories (see Zanardo, 2006) but this would have no substantial bringing for our present purposes.

The relation $\mathcal{M}, (m, h) \models_{Ock} \phi$ expresses the Ockhamist truth-relation between a model, a moment-history pair, and a formula or—as we shall say—'truth in the Ockhamist sense'. The importance of the history-parameter is more sensible when coming to $F\phi$, since in order to evaluate the latter at $(m, h)$, we must so to speak shift forward in time *along the very history h*.

Given an ockhamist model $\mathcal{M}$ we define a corresponding subvaluationist model $\mathcal{M}_{Ab} := (\mathcal{T}, V^+, V^-)$, where $\mathcal{T}$ is as above, $V^+$ (subtruth) and $V^-$ (subfalsehood) are subvaluationist evaluation functions from propositional variables to *sets of moments* which are defined as follows:

$m \in V_{Ab}^+(\phi)$ iff $\exists h(h \in H_m$ and $\mathcal{M}, (m, h) \models_{Ock} \phi)$
$m \in V_{Ab}^-(\phi)$ iff $\exists h(h \in H_m$ and $\mathcal{M}, (m, h) \not\models_{Ock} \phi)$

Truth (subtruth) of a formula in a subvaluationist model ($\mathcal{M}_{Ab} \models_{Ab} \phi$) and subvaluationist validity ($\models_{Ab} \phi$) are defined as usual. The same goes for *logical consequence*: a set $\Delta$ of formulas is a logical consequence of a set $\Gamma$ ($\Gamma \models_{Ab} \Delta$) if and only if whenever *all* formulas in $\Gamma$ are *subtrue* then at least *some* formula in $\Delta$ is *subtrue*. It is then easy to see that:

$\phi, \neg\phi \not\models_{Ab} \phi \wedge \neg\phi$ (failure of Additivity)
$\phi \rightarrow \psi, \phi \not\models_{Ab} \psi$ (failure of MP)
$\phi, \neg\phi \not\models_{Ab} \psi$ (failure of EAQ)

That is, *Ab* is *non-adjunctive*, and *Modus Ponens* (MP) and *Ex Absurdo Quodlibet* (EAQ) fail to be valid rules in it.[7] However, notice that the special cases of single-premise MP and EAQ hold: $\phi \wedge (\phi \rightarrow \psi) \models_{Ab} \psi$ and $\phi \wedge \neg\phi \models_{Ab} \psi$. From the technical point of view, such features suffice to qualify *Ab* as a discussive logic.

In (Ciuni & Proietti, to appear), we prove some remarkable properties of *Ab* and show that single-premise logical consequence in *Ab* coincides with single-premise logical consequence in *Ock* where no modal operator but $P$ and $F$ occurs (Ciuni & Proietti, to appear, Theorem 1)—call such a fragment $Ock^{PF}$.

## 2.1 Proof Methods: A Hint

Is there a sound and complete proof method for *Ab*? Is there a decidable one? How does it work? Provided we have a decidable proof method for classical linear temporal logic—which is the case—we may provide a positive answer to the first two questions and sketch a general strategy—for the purposes of the present work we will not work it out—for the third one.

Suppose we dispose of a terminating algorithm—modal tableaux are the best example (see Mendelsohn & Fitting, 1999 for reference)—for linear time logic (S4.3) which given any finite set $\Sigma$ of formulas

**(a1)** produces a linear temporal model (a branch) for $\Sigma$ if this set is S4.3-consistent

**(a2)** otherwise shows that $\Sigma$ is not satisfiable in any linear temporal model

Then, in order to test if $\Sigma \models_{Ab} \phi$ one must run the above algorithm with the set $\{\psi_i, \neg\phi\}$ for every $\psi_i \in \Sigma$. If the algorithm gives (a2) for *at least one* of these runs, then we say $\phi$ is derivable from $\Sigma$ ($\Sigma \vdash_{Ab} \phi$).

This will constitute a sound and complete proof method. To show soundness, suppose that $\Sigma \not\models_{Ab} \phi$. This amounts to say that we have a model $\mathcal{M}$ which at some $m$ satisfies all $\psi_i$ and does not satisfy $\phi$. Consequently, for

---

[7] The proofs of the two failures are quite elementary, once considered the semantics above. In any case, we refer the reader to (Ciuni & Proietti, to appear).

every $\psi_i$ there is a history $h_i$ such that $\mathcal{M}, (m, h_i) \models_{Ock} \psi_i \wedge \neg\phi$. But since each $h_i$ provides a linear model for $\psi_i$ and $\neg\phi$ the algorithm cannot give (a2) for any of these combinations and then $\Sigma \not\vdash_{Ab} \phi$.

For the completeness part, suppose instead that $\Sigma \not\vdash_{Ab} \phi$. This means that for all sets $\{\psi_i, \neg\phi\}$ the algorithm gives (a1), i.e. we are provided a linear model and a point $x_i \models_{Ock} \psi_i \wedge \neg\phi$. By identifying all the $x_i$ as different couples $(m, h_i)$ of a common root $m$ we can thus build a countermodel which shows that $\Sigma \not\models_{Ab} \phi$.

## 3 Probabilistic Semantics for Future Contingents

There is a fundamental ambiguity between an *objective* and a *subjective* reading of *subtruth*. The first equates subtruth of $\phi$ to the fact that *there are no objective reasons to exclude* $\phi$. Clearly, this is compatible with the lack of objective reasons to exclude $\neg\phi$ as well. Reference here is made to what fails to be *causally determined* by the present state of the world and a reason to endorse this view is given, among others, by a *non epistemic* interpretation of *quantum* properties of our physical world.[8] The second reading equates the subtruth of $\phi$ to the fact that $\phi$ is *supported by someone*—and this may clearly be also the case for $\neg\phi$. In this case we refer to the belief states of agents as they appear in a discussion.[9] Such a reading is closer to the spirit of *discussive logics*, whose aim is indeed modeling a discussion round in terms of the formulas supported by at least one participant.

Both notions have a probabilistic flavour. Indeed it is natural to read the objective notion of subtruth as 'having objective probability higher than 0' and the subjective notion of 'being supported by someone' as 'having an expected chance higher than 0'. Here, we give substance to this probabilistic reading by extending trees with a *probability function B*, which may express both the *objective probability* or *expected chance* that a history takes place.

---

[8] If during an experiment both 'electron will be measured spin up' and 'electron will be measured spin down' are open possibilities at a moment $m$ then there seem to be a reason to hold both possibilities as 'real' objective features of the world at its present state. For a wider analysis of the link between quantum mechanics, branching-time and probability (McCall, 1994).

[9] Thus, the subtruth of 'Tomorrow there will be a sea battle' reduces to the fact that someone supports such a forecast in a discussion, no matter if already dismissed by reality.

## 3.1 Probabilistic Trees

We define *probabilistic trees* $\mathcal{T}^B$ as pairs of the form $(\mathcal{T}, \{B_m \mid m \in T\})$, where $\mathcal{T}$ is a tree and for each moment $m$, $B_m : H_m \longmapsto [0,1]$ is a function from histories to the interval of the reals comprised between 0 and 1. Provided that the set $H_m$ is at most countable we may also assume

(B1) $0 < B_m(h) \leq 1$ for every $h \in H_m$

(B2) $\Sigma_{h \in H_m} B_m(h) = 1$

Here B2 simply states that $B_m$ is a standard probability measure defining the whole fan of possibilities. In the objective reading B1 states that at $m$ any history $h$ in $H_m$ is a 'real' option. In the subjective interpretation B1 reads: at every moment $m$ and history $h$ in $H_m$, is attributed a positive *expected chance*, while B2 states that it is believed that something happens (at least one course of events takes place). Here $B_m$ does not need to represent the expectations of a single agent. In fact $B_m$ may very well be read as the average of the sum of the different agents' expectations, with seems to be sound w.r.t. a discussive interpretation. For the sake of simplicity, we shall call it 'average expectation'. In section 3.3, we will define $B_m$ in terms of assignments of probability by individual agents.

We may also read the function $B_m$ as a probability distribution on the set $\mathfrak{Form}$ of formulas of the language defined as

$$B_m(\phi) := \Sigma_{h_i \in \|\phi\|_m} B_m(h_i)$$

where $\|\phi\|_m := \{h \mid (m,h) \models_{Ock} \phi\}$. It is easy to check that the definition satisfies the *Kolmogorov axioms*, i.e. for any $\phi$ and $\psi$:

**K1** $0 \leq B_m(\phi) \leq 1$

**K2** If $\models_{Ock} \phi$ then $B_m(\phi) = 1$

**K3** If $\phi \models_{Ock} \psi$ then $B_m(\phi) \leq B_m(\psi)$

**K4** If $\phi$ and $\psi$ are mutually exclusive relative to Ockhamist semantics, then $B_m(\phi \vee \psi) = B_m(\phi) + B_m(\psi)$

Finally, we define models and give the truth-clauses for the formulas. Probabilistic tree models $\mathcal{M}^B$ are pairs $(\mathcal{T}^B, V^+, V^-)$, where $\mathcal{T}$ is a probabilistic tree and $V^+$, $V^-$ are the *positive* and *negative* evaluation function, respectively:

$$m \in V^+(\phi) \text{ iff } B_m(\phi) > 0$$
$$m \in V^-(\phi) \text{ iff } B_m(\phi) < 1$$

In the objective reading truth (resp. falsehood) of $\phi$ equates with positive objective probability for $\phi$ (resp. positive probability against $\phi$). Analogously, from the subjective point of view truth of $\phi$ may be read as positive expected chance that $\phi$. We can express truth at a moment $m$ also in the usual way, by defining the truth-relation $\mathcal{M}^B, m \models_B^+$ and the falsehood-relation $\mathcal{M}^B, m \models_B^-$ as follows:

$$\mathcal{M}^B, m \models_B^+ \phi \text{ iff } m \in V^+(\phi) \text{ iff } B_m(\phi) > 0$$
$$\mathcal{M}^B, m \models_B^- \phi \text{ iff } m \in V^-(\phi) \text{ iff } B_m(\phi) < 1$$

Now we briefly explore some relevant features of the semantics and the resulting logic. Clearly, the above semantics is non-recursive, exactly as the subvaluationist semantics presented in (Ciuni & Proietti, to appear). Indeed, there are models $\mathcal{M}^B$, where $\mathcal{M}^B, m \models_B^+ \phi$ and $\mathcal{M}^B, m \models_B^+ \psi$ but $\mathcal{M}^B, m \not\models_B^+ \phi \wedge \psi$. For suppose that $B_m(\phi) = 0.3$, $B_m(\psi) = 0.7$, and $\|\phi\|_m \cap \|\psi\|_m = \emptyset$. This implies that $\phi$ and $\psi$ are supported at $m$, but their conjunction is not. Moreover, the definition of $V^+$ does not exclude that $\phi$ and $\neg\phi$ are both true.[10] This is in accordance with the fact that we may have real opposite possibilities or that, in the subjective reading, mutually exclusive views may be supported in a discussion round.

### 3.2 Equivalence Results

The resulting logic is paraconsistent and non-adjunctive, exactly as $Ab$. Below, we present a more systematic comparison between the notion of logical consequence emerging from the above semantics and the one emerging from the subvaluationist semantics in (Ciuni & Proietti, to appear). For any $B_m$ which satisfies B1 and B2 we have the following:

**Proposition 1** $B_m(\phi) > 0$ iff $\exists h(h \in H_m \text{ and } \mathcal{M}, (m, h) \models_{Ock} \phi)$

*Proof*: The left-right side is easily proved by the contrapositive. Suppose that $\forall h$, if $h \in H_m$, then $\mathcal{M}, (m, h) \not\models_{Ock} \phi$. This implies that $\|\phi\|_m = \emptyset$, and, as a consequence, $B_m(\phi) = 0$. To prove the right-left side, consider

---

[10] If $0 < B_m(\phi) < 1$, the same will hold also for $B_m(\neg\phi)$. Indeed, the basic evaluation is Ockhamist, and thus we have $\|\phi\|_m \cap \|\neg\phi\|_m = \emptyset$ and $\|\phi\|_m \cup \|\neg\phi\|_m = H_m$. By this and $B_m(H_m) = \Sigma_{h \in H_m} B_m(h) = 1$, we have $B_m(\neg\phi) = 1 - B_m(\phi)$. If $0 < B_m(\phi) < 1$, it follows that $0 < B_m(\neg\phi) < 1$.

that if $\exists h(h \in H_m$ and $\mathcal{M}, (m, h) \models_{Ock} \phi)$, then by B1 we have $\exists h(h \in H_m$ and $B_m(h) > 0$ and $\mathcal{M}, (m, h) \models_{Ock} \phi)$. The latter is enough to have $\Sigma_{h_i \in \|\phi\|_m} B_m(h_i) > 0$ and then $B_m(\phi) > 0$. □

The above equivalence is essential to derive the following result showing the extensional equivalence—with respect to the structures which satisfy B1 and B2—of the notion of truth of the subvaluationist semantics for $Ab$ and that of the probabilistic semantics presented here:

**Corollary 2** $m \in V_{Ab}^+(\phi)$ iff $B_m(\phi) > 0$

*Proof*: $m \in V_{Ab}^+(\phi)$ iff $\exists h(h \in H_m$ and $\mathcal{M}, (m, h) \models_{Ock} \phi)$. By this and the above result, it follows $m \in V_{Ab}^+(\phi)$ iff $B_m(\phi) > 0$. □

The notion of *consequence* for $B$ is given by the following:

**Definition 3** $\phi_1, \ldots, \phi_n \models_B \psi$ iff $\forall \mathcal{M}^B, m$ if *for all* $1 \leq i \leq n$ $\mathcal{M}^B, m \models_B^+ \phi_i$, then $\mathcal{M}^B, m \models_B^+ \psi$

Other definitions of consequence are possible for $B$, of course. However, the above definition seems to have an important conceptual feature in its corner. Indeed, it states that a formula is a consequence of some other iff the former is *objectively possible* or *supported* in every model where the latter is. In other words, logical consequence is *preservation* of *objective probability* or, from the subjectivist perspective, of *expected chance*. Thus, our definition seems to capture the general idea of what logical consequence is: preservation of truth, where the specific notion of truth to be considered is the one which is defined by the truth-conditions of the semantics in question.

An immediate consequence of this definition is that the notion of logical consequence defined for $Ab$ coincides with the one just defined for $B$:

**Proposition 4** $\phi_1, \ldots, \phi_n \models_B^+ \psi$ iff $\phi_1, \ldots, \phi_n \models_{Ab}^+ \psi$

*Proof*: For the left-right direction, suppose that $\Phi \models_B^+ \psi$ and $\Phi \not\models_{Ab}^+ \psi$. By Proposition 1, Definition 3 and the definitions of $\mathcal{M}^B$ and $\models_B^+$, the former implies that (1) for every $\mathcal{M}, m$, if for each $\phi_i$ there is a history $h_i$ s.t. $\mathcal{M}, (m, h_i) \models \phi_i$, then there is a history $h'$ s.t. $m \in h'$ and $\mathcal{M}, (m, h') \models \psi$. The latter implies that (2) there are a model $\mathcal{M}_{Ab}$ and moment $m$ s.t. for all $\phi_i$ $\mathcal{M}, m \models_{Ab}^+ \phi_i$ and $\mathcal{M}, m \not\models_{Ab}^+ \psi$. Given the definitions of $\mathcal{M}_{Ab}$ and $\models_{Ab}^+$, (2) contradicts (1) and then the initial hypothesis. The right-left proof proceeds along the same lines. □

A remarkable consequence of the result above is that all the failures of classical inferences in $Ab$ transmits to $B$. Indeed, when it comes to the rules of inference EAQ does not hold: $\phi, \neg \phi \not\models_B^+ \varnothing$. The same happens to adjunctivity: $\phi, \neg \phi \not\models_B^+ \phi \wedge \neg \phi$.[11] Proposition 4 guarantees also other failures, some valid inferences and—more in general—important features of $B$. Clearly, it implies the failure of multi-premise MP. Conversely, the fact that single-premise MP holds in $Ab$ is enough to secure that it holds in $B$. More important, Proposition 4 is enough to prove that: ($\star$) single-premise consequence in $B$ coincides with single-premise consequence in standard linear-time logic. This fact is a consequence of Proposition 4 and Theorem 1 in (Ciuni & Proietti, to appear), which we quote here for reasons of convenience:

**Proposition 5** (*Ciuni & Proietti, to appear, Theorem 1*)  For any formulas $\phi, \psi_1, \ldots, \psi_k$ of the purely temporal language (i.e. without additional alethic operators $\Box, \Diamond$) the following equivalence holds:

$$\phi \models_{Ock} \psi_1, \ldots, \psi_k \text{ iff } \phi \models_{Ab} \psi_1, \ldots, \psi_k$$

The theorem proves that single-premise consequence in $Ab$ is extensionally equivalent with single-premise consequence in that fragment of Ockhamist logic which does not contain occurrences of $\Box$ (section 2 is enough to understand that the result *does not* extend to multi-premise consequence). Since consequence in that fragment of Ockhamist logic is extensionally equivalent to consequence in standard linear-time logic, by Theorem 5 and Proposition 4, we get ($\star$). This is very interesting, since it tells us that, when just one premise is into account, reasoning in $B$ does not differ from the standard reasoning in linear-time logic—which is taken to capture the most basic inferential procedures about temporal statements. In other words, our probabilistic reasoning is perfectly acceptable from the standard point of view on temporal reasoning.

Considered together, Proposition 1, Corollary 2 and Proposition 3 prove that the two semantics $Ab$ and $B$, though encoding different readings of truth, *generate the same logic* or—in other terms—are *invariant* w.r.t. inferential procedures and validity. As a consequence, the discussive logic resulting from the subvaluationist semantics in (Ciuni & Proietti, to appear) may be given a probabilistic reading.

---

[11] Indeed, it can be the case that both $\phi$ and $\neg \phi$ are supported in a discussion round, and yet no one supports $\phi \wedge \neg \phi$, by the very definition of $V$—this feature is in accordance with the fact that we are presupposing *rational* agents, that is agents who do not believe contradictions.

## 3.3 The Source of Average Assignments of Probabilities

In the previous section, we suggested that subjective truth may be read as 'average expectation', i.e. the average of the sum of the different agents' expectations. Thus, the notion does not express any specific individual probability assignment by agents. However, it presupposes them. We may indeed define *average expectations* in terms of *individual agents* and *individual probability assignments*. This will in turn explain the *source* of the average expectation which we introduced earlier. First, we extend our *probabilistic trees* $\mathcal{T}^B$ into *augmented probabilistic trees* $\mathcal{T}^{B^+}$ $(\mathcal{T}, \{B_m^{a_i} \mid m \in T \text{ and } a_i \in Ags\})$, where $Ags$ is a countable set $\{a_1, \ldots, a_n\}$ of agents. Second, for every $h$, we define $B_m$ as follows:

$$B_m(h) = \frac{B_m^{a_1}(h) + \ldots + B_m^{a_n}(h)}{n}$$

In other words, for every history, the probability assignment $B_m$ we have introduced in the previous section obtains by averaging out the probability assignments $B_m^{a_i}$ of each individual agent $a_i$ to that history. It just takes elementary mathematics to prove that this is still a well-behaved probability distribution, i.e. it respects B2, and that

$$B_m(h) > 0 \text{ iff } B_m^{a_i}(h) > 0 \text{ for at least some agent } a_i$$

This tells us that every agent plays a role in computing the average expectation, and the latter cannot obliterate the probability assignment of any single agent. In determining the average expectations, all agents are *equal*—by this we mean that no given agent has a special role or weight in the way $B_m(h)$ is established, and the way the latter is computed cannot make the assignment of some agent non-influential. By the definition of $B_m(\phi)$, it is trivial to prove that

$$B_m(\phi) = \frac{\Sigma_{h \in \|\phi\|_m} B_m^{a_1}(h) + \ldots + \Sigma_{h \in \|\phi\|_m} B_m^{a_n}(h)}{n}$$

or, equivalently,

$$B_m(\phi) = \frac{B_m^{a_1}(\phi) + \ldots + B_m^{a_n}(\phi)}{n}$$

As above, we have that $B_m(\phi) > 0$ iff $B_m^{a_i}(\phi) > 0$ for at least some agent $a_i$. Once again, this tells us that agents are *equal* in determining the average expectation on $\phi$, in the sense that the way $B_m(\phi)$ is computed does not attribute any special role to any agent and cannot obliterate the contribution of any of them.

## 4 Discussion and Conclusions

In the present paper we have solved the ambiguity between an *objective* and a *subjective* reading of the paraconsistent logic proposed in (Ciuni & Proietti, to appear), gave a new probabilistic semantics and explained how this fares with them. The last move has allowed us to establish the formal results in section 3 and to prove that the *subvaluationist* semantics of (Ciuni & Proietti, to appear) is, under certain conditions, equivalent with the *probabilistic* one introduced here. Finally, we clarified the subjective reading by grounding the notion of 'average expectation' on the individual expectations of the agents in question.

Coming to the general merit of our present proposal, one might wonder why we should combine discussive logic with future contingents. One reason lies in the subjective reading and is connected to situations where discussions precede a *deliberation*—we think of assemblies, parliaments, councils. When it comes to such situations, discussions about the future play a prominent role, since the orientation toward a given deliberation or vote is clearly influenced by the different beliefs about what will happen. For instance, if a member of a given parliament believes that the international relationships will become increasingly more tensed, she will vote—likely— for the increase of military expense, while another member will vote against increasing it, if she believes there will not be a growing tension.

Our logic mirrors what is given expression in discussion rounds of the sort above, though it cannot mirror the fact that the majority of the agents support a view rather than another, or other quantitative criteria in the procedure of deliberating (deliberation by unanimity, etc.). The importance of the logic lies in the fact that it can represent the plurality of views which motivate and substantiate any *democratic* procedure of deliberation, independently of the concrete mechanism which computes the winning option and the loosing ones. Appeal to 'democratic procedures' here is due to the fact that our apparatus models a society of *equals*, where the computation of the average expectations *cannot obliterate* the individual expectations of any individual agent. In other words, in our apparatus no supported view can fail to be recorded in the debate emerging from the discussion round:

any participant is given a voice, and all voices have the same weight.

Before closing, we focus on a technical issue. An interesting feature of our semantic apparatus is that it is not *committed* with either paraconsistency or a notion of 'true' as 'not completely uncertain in the agents' expectations'; on the contrary, it proves to be a very plastic tool. A stricter notion of truth obtains if we impose the value of $0,5$ as a (loose) *threshold*: in this case, $\phi$ would be true at $m$ iff $B_m(\phi) \geq 0,5$. '$\phi$ is true' would then read '$\phi$ is *equally or more* supported that $\neg \phi$'.[12] 'Playing' with the threshold also allows to get a *non-paraconsistent* notion of 'average expectations': it suffices to impose $B_m(\phi) > 0,5$ as the threshold for $\phi$ to be true (at a given moment $m$).[13] Our apparatus can thus be disentangled from the application to paraconsistency and find a wider scope of applications.

We conclude by considering some possible future developments. The setting we propose represents discussions where any logically consistent view may be supported, independently from its (objective) *plausibility*. As a consequence, if 'The moon is made of cheese' is supported in a given discussion round, our logic will not be able to describe the fact that a basic evolution of our knowledge of the world will result in abandoning that view. Also, the setting is able to model a discussion round, but it cannot model that *change* of views which naturally results from a discussion, due to the exchange of opinions. These are in turn serious limitations, since we want to discharge blatantly implausible views *via* the update of our information on the world, and since –more in general– our discussions also aim to *refine* our views –ideally, *via* the announcement of some information– rather than to produce a stubborn repetition of our views. We believe that such limitations may be overcome by extending our logic with a *modal update operator* in the style of Dynamic Epistemic Logics.[14] We wish to do that in future research.

---

[12] Notice that this would not change the logic (that is the class of valid formulas and the relation of logical consequence). The reason for such a threshold would then be conceptual, and connected to the need of a stricter notion of truth.

[13] A consequence of this move is that $\phi$ and $\neg \phi$ cannot both be true in the same model at the same moment (since if $B_m(\phi) = 0.51$, then $B_m(\neg \phi) = 1 - B_m(\phi) = 0.49$, due to axiom **K4** in section 3) and EAQ is valid (since no model of the logic would verify both $\phi$ and $\neg \phi$ at the same moment). It is easy to check that Additivity and MP would still fail, though. Also, notice that the resulting logic would *not* be *paracomplete*, since $\phi \vee \neg \phi$ would be valid in it, due to axiom **K4** and the definition of the threshold.

[14] We think especially of operators for *updates* and *public announcements*, which are proposed by many different works in the tradition of Dynamic Epistemic Logics.

## References

Baier, C., & Kwiatkowska, M. (1998). Model checking for a probabilistic branching time logic with fairness. *Distributed Computing, 11*, 125–155.
Ciuni, R., & Proietti, C. (to appear). The abundance of the future. A paraconsistent approach to future contingents. *Logic and Logical Philosophy*.
Hasle, P., & Øhrstrom, P. (1995). *Temporal logic from ancient ideas to artificial intelligence*. Kluwer.
Jaśkowski, S. (1969). Propositional calculus for contradictory deductive systems. *Studia Logica, 24*, 143–160.
McCall, S. (1994). *A model of the universe. Space-time, probability, and decision*. Oxford: Clarendon Press.
Mendelsohn, R., & Fitting, M. (1999). *First-order modal logic*. Dordrecht: Kluwer Academic Publishers.
Thomason, R. (1970). Indeterministic time and truth-value gaps. *Theoria, 36*, 264–281.
Zanardo, A. (2006). Quantification over sets of possible worlds in branching-time semantics. *Studia Logica, 82*, 379–400.

Roberto Ciuni
Institut für Philosophie, Ruhr-Universität Bochum
GA 3/39, Universitätsstraße 150, D-44780 Bochum, Germany
e-mail: ciuniroberto@yahoo.it
URL: http://www.ruhr-uni-bochum.de/philosophy/logic/team/ciuni.html.en

Carlo Proietti
Department of Philosophy, University of Lund
Kungshuset, Lundagård, 222 22 Lund, Sweden
e-mail: Carlo.Proietti@fil.lu.se
URL: http://www.ht.lu.se/o.o.i.s?id=23921p=Carlo Proietti

# Overview of a Tarskian Solution to the Iterated Prisoner's Dilemma

## Christopher N. Foster

**Abstract:** The iterated prisoners dilemma backwards induction paradox is one of the more intractable problems of game theory. The majority opinion, based on backwards induction, is that universal defection is uniquely rational in finite game cases. I argue that because cooperation has superior results, universal defection cannot be rational. The challenge is to find the flaw in the backwards induction argument. My approach shows that if we make a Tarskian distinction between object-rationality and meta-rationality we can show that conditional cooperation is rational after all. The meta-rational agent chooses that strategy such that, if the other player responds rationally, one is as well off as possible. Optimal meta-strategies involve conditional cooperation, incentivizing the other player to cooperate until the final round.

**Keywords:** prisoners' dilemma, meta-rationality, iterated, repeated, paradox, Tarski, logic, rationality, game theory, defection, dominance principle, cooperation

The prisoner's dilemma presents a familiar two-player problem in which individual rationality results in poor results for each. While the result of the prisoner's dilemma is sub-optimal, it is not quite paradoxical. We discover a much more troubling problem, however, in the case of the *iterated* prisoner's dilemma. It would seem obvious that if two rational agents are going to play many games in a row then cooperation would be highly mutually beneficial. However, it has been established by backwards induction that rational agents will actually defect in every round, even if it results in terrible outcomes for both. Most game theorists simply accept this outcome, difficult though it may seem, reasoning that it is what logic dictates. Others seek to justify cooperative approaches despite the conclusions of logic or by altering the assumptions (e.g. Bovens, 1997; Kreps, Milgrom, Roberts, & Wilson, 1982; Binmore, 1997). This paper outlines a resolution to this paradox, analogous to Alfred Tarski's answer to the liar, by exposing a subtle error within the assumptions of the backwards induction argument, thus paving the way for rational agents to enjoy the benefits of mutual cooperation without ignoring logic.

Before offering my response, I begin by presenting the argument for mutual defection. Suppose the prisoner's dilemma set up is as follows:

|  | Player B Cooperates | Player B Defects |
| --- | --- | --- |
| Player A Cooperates | A: 1, B: 1 | A: -2, B: 2 |
| Player A Defects | A: 2, B: -2 | A: -1, B: -1 |

According to standard reasoning, in both players will follow the *dominance principle* and defect. According to the dominance principle (Nozick, 1969), since, no matter what the other player does, each player is better off defecting, it follows that both players should defect. This is the commonly accepted outcome of the single-game prisoner's dilemma (Olin, 1988).

This outcome is unfortunate because it is worse for both players than mutual cooperation, but it is not quite paradoxical. The situation is much worse in the case of the *iterated* prisoner's dilemma. If both players know that they will play exactly one hundred games then a familiar backwards induction argument shows that the two should defect the entire time. The argument states that at game one hundred, since both players know that this is the final game both will treat it as a single game and defect. Since both players realize this inevitable outcome in the final game, both will then also treat the $99^{th}$ game the same as the single game case (since their choices will not affect the result of the final game) and mutually defect. In the $98^{th}$ game both players will reason that since the outcome of the final two games is unalterable, both should defect here too. The reasoning continues all the way to the first game, resulting in mutual universal defection (MUD) (Kreps et al., 1982).

Many economists, philosophers, and logicians simply accept this result as the inevitable conclusion of logic, seeing cooperation at any stage of the game as proof of irrationality on the part of the cooperating player. While a final score of -100 each is terrible, there simply seems to be no way around it if both players are rational. My paper presents a three-part argument that this conclusion is not logically acceptable and that rational agents should cooperate after all.

The first step of my argument demonstrates that the iterated prisoner's dilemma is not just unfortunate but a genuine paradox by showing that constant mutual defection is *also* logically unacceptable. If logic entails contradictory conclusions then the challenge is not simply to 'bite the bullet' but rather to uncover the hidden error within the assumptions that leads to the contradiction. The second step shows which assumption from the back-

# Overview of a Tarskian solution 17

wards induction argument to reject and why. The final stage demonstrates the solution to the dilemma that follows from these insights.

That the problem is paradoxical is demonstrated by showing that it leads to one to be rational by being *irrational* (just as the liar sentence is true by virtue of being false), and vice-versa. Pettit and Sugden (1989) argue that it would be wiser for a player to begin by *cooperating* in order to throw off the other player's assumption of his/her rationality. If the other player no longer believes the first player to be rational, the second player may then realize that conditional cooperation is advantageous after all, resulting in many rounds of mutually beneficial cooperation.

Their result, however, reveals a contradiction: If being *irrational* results in a better outcome then it turns out to be *rational* after all! Pettit and Sugden's results contradict their own terms, but the way they do so reveals a difference between object-rationality (that is assumed as part of the situation) and *meta-rationality* (with which one acts upon that situation). Petit and Sugden's player is *meta-rational* by virtue of being *object-irrational*. This paper argues that equivocation between the two leads to paradox just as equivocation between levels of truth leads to the paradox of the liar. The claim of this paper is that rationality is a *meta*-predicate in the Tarskian sense (Tarski, 1936). This insight reveals the faulty assumption that leads to the paradox: Just as a Tarskian truth predicate is not allowed to occur within the language it modifies, so the assumption of one's own rationality cannot be part of the situation upon which one acts rationally, on pain of contradiction.

A simple example demonstrates this insight. If one's own rationality can be part of the situation upon which one is supposed to act then we can generate a paradox as follows: Suppose you are offered a choice of box A, which contains $10, or box B, which contains nothing. However, you are also informed that you will be given the sum of $100 if you choose *irrationally* (I am supposing that if one's rationality can be part of the situation then so can one's irrationality). What is one to do? One would seem tempted to choose box B in order to cash in on the $100 bonus. However, if that strategy works then it would be *rational after all* (because it results in more money), and therefore it should *not* result in the bonus. One would then consider choosing box A (in order to be irrational), but if this strategy worked (by gaining the bonus) then that would make it rational after all, so it shouldn't work. We have the game-theoretic equivalent of the liar's paradox: One is rational in this situation if and only if one is irrational.

The Tarskian insight teaches us that there is a subtle equivocation going on between the rationality that is part of the situation (object-rationality—

or *irrationality* in this case), and the rationality with which one contemplates how to act upon the situation (*meta*-rationality). The equivocation between the two explains how a simple game-theoretic situation can lead to the contradictory conclusion that one is rational if and only if one is irrational. When the equivocation is resolved we arrive at the non-contradictory insight that in this situation one is meta-rational just in case one is object-irrational; the paradox is resolved. If this is right then, just as Tarski's truth predicate may not be part of the language it modifies, the assumption of one's own rationality may not be part of the situation upon which one is supposed to act rationality.

There is no contradiction, however, in assuming the rationality of the other player and acting rationally upon such a situation. However, our situation also cannot include the assumption that the other player *knows* one's own rationality (for that would again make one's own rationality part of the object-situation). One simply treats the other player's rationality as part of the object-situation upon which one must act from a *meta* point of view. If we revisit the repeated prisoner's dilemma with the revised set if assumptions that the other player is object-rational but not that one's self is, we find the most rational meta-strategy to employ: It is to *choose the strategy that, if the other player responds rationally, will leave one's self the best off*. Any strategy that fits this bill will involve conditional cooperation. One cooperates in order to incentivize the other player to cooperate as well. *Tit-for-tat* is just such an optimal meta-rational strategy (its benefits are demonstrated, for example, in Axelrod, 1984).

There is a strange consequence of the meta-rational approach: It directs one to cooperate even in the *final round*. If one were not going to cooperate (conditionally) in the final round then there would be no incentive for the other player to cooperate in the previous round. Yet there seems to be no way for the other player to *know* that one is planning to cooperate conditionally in the final round; therefore there is no way for the other player to know that he or she has an incentive to cooperate in the previous round.

Even if conditional cooperation has much better results than MUD, one still wonders how the other player is to *know* whether one intends to cooperate in the final round. The simple answer is that he or she *doesn't*; he or she simply has to *guess* one's strategy and to respond as well as he or she can. It might then be tempting to defect in the final round (to gain that extra point), but, this could lead to a backwards induction slippery slope, all the way back to MUD. The meta-rational approach works by creating the *incentive* for the other player to cooperate as long as possible. The meta-

## Overview of a Tarskian solution

rational strategy in effect *makes it rational* for the other player to cooperate prior to the final round. If one plays tit-for-tat and the other player defects in the penultimate round then the other player ends up with a score of only 99 (rather than 101 if he or she had waited until the final round). A player in fact loses two points for every round of premature defection. A player that defects too early therefore has chosen *irrationally* (relative to one's chosen meta-strategy), knowingly or not. An area for further research is the sense in which *unknown* incentives can affect the rationality of an action (Brams (1975), allows players to *announce* their meta-strategies so that the other player will know them).

We still have a paradox on our hands, but we are in good company. The greatest paradoxes seem never to be finally resolvable but remain to taunt us even after hearing our preferred solutions. My approach leaves one player in defiance of the seemingly valid dominance principle and seems to suppose that the other player can respond rationally to one's strategy without knowing exactly what it is. However, as with being a one-boxer in Newcomb's problem, my approach has results to advocate for it: One who chooses a meta-rational strategy (assuming the other player responds rationally) receives a final score of 97, while one who follows the dominance principle (with backwards induction) receives an abysmal -100. As with Newcomb's one-boxers, I would rather be laughed at for being 'irrational' all the way to the bank!

Another potentially problematic observation is that the two players are in symmetrical situations, so why doesn't the other player choose meta-rationality, allowing one to defect in the final round (and get the 101 rather than 99)? My answer is that it is indeed possible that the other player may choose the same meta-strategy (resulting in a happy 100, 100). I do not take this possibility to represent an objection to my approach. Since neither player can see the other's strategy in advance, both players are in a guessing game as to the other's late-game strategy (resulting in multiple possible end games, based upon the other player's choices). What is relevant here, however, is that *if* one attempts to out-defect the other, one runs the risk of defecting too early and missing out on any remaining rounds of potential cooperation. In this guessing game one is advised to err on the side of caution since every round in which one defects too early costs two points, whereas every round guessed late (allowing a tie rather than defecting first, or allowing the other to defect first rather than a tie) only costs one point. Therefore, the expected utility of both players is maximized by postponing defection as long as possible.

Though the problem of final-round cooperation and asymmetry are indeed strange results, they follow from the Tarskian approach and constitute much better bullets to bite than the results of mutual universal defection. These problems and their answers will be addressed more fully in the complete article (in preparation). My goal here is not to show that this paradox no longer *paradoxical* but to explain why cooperation is actually the most rational strategy (rather than MUD). This is accomplished by demonstrating that adopting the meta-rational perspective breaks the symmetry between the two players and prevents them from outguessing each other to oblivion. In this way the backwards induction is stopped, and the two players can work from an expected utility perspective, rather than a dominance perspective, and thereby enjoy many rounds of fruitful cooperation.

In conclusion, in the paradox of the iterated prisoners dilemma, as in Tarski's liar, a contradiction is generated due to a hidden error within the assumptions, namely, including a meta-predicate within the object situation it is supposed to modify. Removing that predicate and banishing it to the meta-sphere results in rewarding solutions to some of history's most daunting paradoxes. In the case of the prisoner's dilemma the intuition that mutual cooperation is more sensible than universal defection (since it results in better outcomes) turns out to be logically justified after all.

## References

Axelrod, R. (1984). *The evolution of cooperation*. New York: Basic Books.

Binmore, K. (1997). Rationality and backward induction. *Journal of Economic Methodology, 4*, 23–41.

Bovens, L. (1997). The backward induction argument for the finite iterated prisoner's dilemmas and the surprise exam paradox. *Analysis, 57*, 179–186.

Brams, S. J. (1975). Newcomb's problem and prisoners' dilemma. *The Journal of Conflict Resolution, 19*, 596–612.

Kreps, D., Milgrom, P., Roberts, J., & Wilson, R. (1982). Rational cooperation in the finitely repeated prisoner's dilemma. *Journal of Economic Theory, 27*, 245–252.

Nozick, R. (1969). Newcomb's problem and two principles of choice. In Campbell & Sowden (Eds.), *Paradoxes of rationality and coopera-*

*tion: Prisoner's dilemma and Newcomb's problem.* (pp. 107–133). Vancouver: The University of British Columbia Press.

Olin, D. (1988). Predictions, intentions, and the prisoner's dilemma. *Philosophical Quarterly, 38*, 111–116.

Pettit, P., & Sugden, R. (1989). The backward induction paradox. *Journal of Philosophy, 86*, 169–182.

Tarski, A. (1936). The concept of truth in formalized languages. In Tarski (Ed.), *Logic, semantics, metamathematics: Papers from 1923 to 1938, second edition* (pp. 152–278). Indianapolis: Hackett Publishing Company.

Christopher N. Foster
Department of Philosophy
Utah Valley University
800 West University Parkway
Orem, UT 84058, USA
e-mail: `fosterch5@gmail.com`

# Axiomatising Questions

## CHRIS FOX

**Abstract:** Accounts of the formal semantics of natural language often adopt a pre-existing framework. Such formalisations rely upon informal narrative to explain the intended interpretation of an expression—an expression that may have different interpretations in different circumstances, and may supports patterns of behaviour that exceed what is intended. This ought to make us question the sense in which such formalisations capture our intuitions about semantic behaviour. In the case of theories of questions and answers, a question might be interpreted as a set (of possible propositional answers), or as a function (that yields a proposition given a term that is intended to be interpreted as a phrasal answer), but the formal theory itself provides no means of distinguishing such sets and functions from other cases where they are not intended to represent questions, or their answers. Here we sketch an alternative approach to formalisation a theory of questions and answers that aims to be sensitive to such ontological considerations.

**Keywords:** questions, answers, ontology, methodology

## 1 Introduction

We first introduce some issues and concerns relating to the semantic analysis of questions and answers that are raised, or alluded to in the literature. We then summarise, in broad terms, existing formal analyses. Here we focus on semantic concerns relating to the analysis of direct questions and their answers, rather than pragmatic issues, indirect questions, or the analysis of discourse.

### 1.1 Questions

There are various types of questions that can be posed, even discounting implicit or indirect questions. These can be broadly classified as "wh-questions"—*"Who went to London?"*, *"What was his name?"*, *"How did you do that?"*—"polarity" questions—*"Do you like cheese?"*, and "choice" questions—*"Do you want tea or coffee?"*. The latter might be seen as a variant of a polarity question where there is a forced choice.

Questions can also be embedded in other propositions, for example in certain kinds of propositional attitudes—*"He knows who ate all the cheese"*, *"He wonders whether John likes Mary"*.

Questions may be refined by an additional question—*"Who ate the cheese, was it John?"*, *"Who ate the cheese, or was it the pickle?"*.[1]

## 1.2 Answers

When questions are answered directly, the answer can be presented as a constituent—Q: *"Who went to London?"* A: *"Peter and Mary"/"Nobody"*, or a proposition—A: *"Peter and Mary went to London"/"Nobody went to London"*. Answers may also be indirect, requiring some reasoning to deduce the intended answer.

Polarity questions can be answer with "yes" or "no", an adverb (of an appropriate nature) or a modal expression—Q: *"Do you like cheese?"*, A: *"Yes"/"Sometimes"*. The related proposition may also be spelt out in full—A: *"Yes, I like cheese"/"I like cheese"*.

Choice offering questions are answer with the respondent's choice being identified—Q: *"Do you want tea or coffee?"*, A: *"Tea"/"I would like tea, please"*.

Answers are typically incomplete and non-exhaustive. The *appropriate* answers depend upon the context. What counts as a *relevant* answer may also depend upon what is assumed to be known.

A question may also be answered with a question—Q: *"Who ate the cheese?"*, A: *"Who do you think?"/"How should I know?"*.[2]

## 1.3 Relationship between Focus and Answers

The topic–focus contrast (see Jackendoff, 1972; von Stechow, 1981, for example) is relevant when considering the interpretations of questions and answers (Krifka, 2001; von Stechow, 1991). The *focus* of an utterance may be marked by some form of prosodic emphasis in speech. It typically draws attention to new information that is being introduced, or is in question. This is in contrast to the *topic*, which is taken to be understood, or presupposed. In the case of written language, we may use other cues to deduce what is in focus.

Questions provide a diagnostic test for focus: we can determine what is in focus in a proposition by considering what question the proposition might answer (Paul, 1880; Rooth, 1992)—*"**John** likes cheese"* (Q: Who

---

[1] This could be viewed as a simple example of extended discourse relating to a "question under discussion" (Ginzburg & Sag, 2000; Roberts, 1996; Stalnaker, 1978).

[2] As when clarifying a question with an additional question, this may be an example where a more general analysis of the pragmatics of discourse is appropriate.

*likes cheese?*), *"John likes **cheese**"* (Q: *What does John like?*). Given the nature of these diagnostic tests, it is tempting to argue that wh-terms in questions correspond to the expected focus in their answers, and that it may be appropriate to consider them as either in focus, or at least a place-holder for focus.

Focus, or emphasis, in a question may also clarify what information is sought in the case of polarity questions. In particular, knowledge of what is in focus may assist the addressee in identifying helpful information in the event of a "no" answer—Q: *"Did **John** eat the cheese?"*, A: *"No, it was Peter"* (cf. *"No, it was the banana"*).[3]

For wh-questions, focus may emphasise exactly what is in question (and thus what would constitute an appropriate answer)—*"Who **ate** the cheese?"*, *"Who ate the **cheese**?"*, *"**Who** ate the cheese?"*.

## 1.4 Semantic Theories of Questions

Very broadly, the various formal semantic analyses of questions and their answers can be consider to fall broadly into two camps, namely *questions as sets of answers* and *questions as structured meanings*.[4] There are some common issues for both approaches, such as whether or not questions, answers, and propositions should all be considered to be essentially the same kind of thing.

*Questions as sets of answers*

On the first approach, questions are conceived as sets of (possible) answers (Hamblin, 1958, 1973; Karttunen, 1977, and others). Within the the possible worlds framework, this can be formulated in terms of partitions of worlds (Groenendijk & Stokhof, 1984, 1997), where each partition represents a proposition/answer, and true answers are those partitions that contain the current world.

This approach directly models full, propositional answers. It also allows for some forms of "indirect" answer. Modelling constituent answers require a bit more work, for example by combining the constituent answer with an appropriate abstract derived from the question to produce a full proposi-

---

[3]This is rather like a wh-question with a follow-up question (section 1.1, *"Who ate the cheese, was it John?"*). Such similarities merit attention in any comprehensive account of questions and answers in dialogue.

[4]Some seek to bridge this gap (for example, Aloni & van Rooy, 2002).

tional answer. The analysis of questions and answerhood is in effect given in the same terms as the notion of truth for propositions.

Arguably, issues concerning topic–focus, and how they relate to answers-questions, are not so easily accounted for. There are also other difficulties concerning "goodness of fit" with the data, including the the handling of choice-offering questions (Krifka, 2001).

*Questions as structured meanings*

The structured-meanings account of questions, and answers, can be motivated by observing that the topic–focus may be analysed in terms of such structured meanings (Ginzburg, 1992; Ginzburg & Sag, 2000; Halliday, 1967; Hull, 1975; Krifka, 1991, 2001; Tichý, 1978; Vallduví, 1992, 1993; von Stechow, 1982, 1991; von Stechow & Zimmermann, 1984) in which "old" information is distinguished from "new" information. If wh-terms are considered to be focus-like, then we can take a similar approach to questions. The focus of the answer is corresponds to the "new" information being sought by the question. Constituent answers are answers in which the topic has been elided.

Structured meanings can be considered as "pairs" for both questions—where a wh-term is paired with the body of the question—and for topic-focus structures—where the focus terms is paired with the topic (cf. Krifka, 2001). The "focus" of an answer can be thought of as providing a "filler" for the wh-term. There is a sense in which the wh-term is "abstracted" out of the body of the question. In some accounts, a question is overtly represented by a (typed) $\lambda$-abstract, $\lambda x_T.p$, where $x_T$ corresponds to the wh-term that is abstracted from the body of the question, $p$ (cf. Ginzburg & Sag, 2000; Hausser, 1983; Hausser & Zaefferer, 1979; von Stechow & Zimmermann, 1984). Constituent answers (or focus-terms) can then be "applied" to the question (or fill in the "missing" focus) to give a propositional answer.

Such an approach can give a straight-forward account of constituent answers, although dealing with propositional answers requires some additional formal machinery. The structured-meanings account goes someway towards relating the notion of topic–focus with that of question–answer. It can also provide accounts of choice offering questions (Krifka, 2001). Arguments can also be made that it is not as reductive as the "sets of answers" account.

## 2 Ontological Issues

There are a number of grounds for evaluating the pros and cons of different approaches[5], including coverage of the data, and sensitivity to ontological issues. It is the latter that we focus on here.

In the case of set-theoretic accounts (section 1.4), is it right to reduce all ontological notions to those of sets, or to possible worlds and relations over possible worlds, construed as sets?

We might consider arguments from (Benacerraf, 1965) on numbers, where the critical point is that numbers have structural properties whose status is independent of any particular set-theoretic characterisation. And different set-theoretic characterisations give rise to different unintended consequences—consequences that are not in accord with our understanding of the notion of number.[6] We may wonder whether semantic notions, such as *questions* and *answers* should also be considered to have structural properties that are independent of a set-theoretic characterisation (Fox & Turner, 2012).

It seems odd for questions to *be* their (possible) answers. If that were the case, then any set of propositions/worlds would be a question. This gives rise to a methodological conundrum: how can we explore the relationship between the meaning of a question and its possible answers if a question actually *is* its answers, as assumed by those that follow Hamblin (1973)? It seems natural to argue that we can consider questions and their formal properties without being obliged to engage in some form of ontological reduction, just as we do with numbers. But a set-theoretic reduction appears to rule this out, at least in the formalisation.[7]

Similar arguments apply to accounts that use structured meanings to represent questions. For example, if structured meanings are pairs, how are such pairs to be distinguished from other pairs? And if questions are abstracts, how are such abstracts distinguished from other abstracts/functions?

For some, such reductions may be considered desirable. For example,

---

[5] For example, see (Krifka, 2001) for criticism of the sets-of-answers approach.

[6] Some set-theoretic representations allow us to express seemingly incoherent statements such as "$2 \in 3$". Whether such a statement is true or false does not reflect any intuitions about numbers themselves but is merely a contingent artifact of the chosen representation.

[7] It may be possible to defend a set-theoretic reduction, and say that to entertain a question and its possible answers is to reflect on a membership relationship, or the extensional identity of a question. But then we risk stumbling into a version of the paradox of analysis (Black, 1944). We surely want to say that "considering a question and its answers" means something different to "considering the membership of a set".

Hamblin (1973); Karttunen (1977); Tichý (1978), and others, argue that questions and propositions should be the same type and that any distinction resides in our relationship to them. But this does not necessarily avoid the problem. Even if the notion of being a question is a relational one, then it can still be characterised. And presumably we wish to characterise it in a way that does not allow it to be conflated with some other, fundamentally distinct notion.

To summarise: with set-based accounts, any intuitive, ontological distinction between questions and answers, and arbitrary sets of propositions, is lost; and with conventional structured-meaning accounts, the intuitive, ontological distinction between questions (and answers), pairs, or propositional abstracts, is also lost.[8]

## 3 Towards a Non-reductive Analysis

Here we demonstrate a framework in which a non-reductive theory of questions can be developed. Questions will be treated somewhat like "specifications" in computer science. Answers will be those things that may "satisfy" a specification. Full propositional answers are taken to have a focus that satisfies the specification. Polarity questions will be treated as questions that are answered by a propositional operator of some kind (modal or adverbial).

For our current purpose, it is not essential for us to spell out all the details here; the key objective is to demonstrate that no ontological reduction is required in order for us to develop a formal semantic analysis. We will illustrate this approach using Typed Predicate Logic (TPL), a generic framework of types and predicate logic (Turner, 2008, 2009), described below (section 3.1). TPL frees us from the formal constraints and ontological commitments of other more rigid frameworks. In particular, it allows us to incorporate aspects of the intended interpretation into the formalisation itself, in the form of judgements and types. Some of the details of the formalisation follows the spirit of the structured-meaning approach to questions (section 1.4).[9]

---

[8] We could parody this reduction by observing that there appear to be no questions or answers in these formal theories of questions and answers.

[9] The structured-meanings perspective seems better suited to our aim of avoiding ontological reductions. It may be possible to give a non-reductive account based on some of the insights of the set-theoretic approach. We could seek to define an answerhood relation that correspond to an answer "belonging to" a question. We might hope that this would turn out to be consistent with a non-reductive theory that takes structured meanings as a starting point.

## 3.1 Typed Predicate Logic (TPL)

Typed Predicate Logic is a framework in which various kinds of theories can be formulated, both their "syntax" and proof theory. There are four basic judgements.

$$
\begin{array}{ll}
t \text{ Type} & t \text{ is a type} \\
s : t & s \text{ belongs to type } t \\
t \text{ Prop} & t \text{ is a proposition} \\
t \text{ (or } t \text{ True)} & t \text{ is true}
\end{array}
$$

Theories can be formulated in this system using sequent style rules. We use a context ($\Gamma$) to simplify the presentation of rules that involve discharged assumptions. Formation rules are used to specify the grammar of a theory. For example, we can give the formation rules for conjunction and negation in a propositional logic.

$$\frac{\Gamma \vdash s \text{ Prop} \quad \Gamma \vdash t \text{ Prop}}{\Gamma \vdash (s \wedge t) \text{ Prop}} \wedge^F \qquad \frac{\Gamma \vdash t \text{ Prop}}{\Gamma \vdash \neg t \text{ Prop}} \neg^F$$

Rules governing the "logical" behaviour of such expressions can be given in terms of judgements of truth. These are formulated with constraints that ensure they only apply to the appropriate kinds of entities. We can exemplify this with introduction and elimination rules for conjunction and negation of propositions.

$$\frac{\Gamma \vdash s \text{ Prop} \quad \Gamma \vdash t \text{ Prop} \quad \Gamma \vdash s \quad \Gamma \vdash t}{\Gamma \vdash s \wedge t} \wedge^+ \qquad \frac{\Gamma \vdash s \text{ Prop} \quad \Gamma, s \vdash \bot}{\Gamma \vdash \neg s} \neg^+$$

$$\frac{\Gamma \vdash s \text{ Prop} \quad \Gamma \vdash \text{ Prop} \quad \Gamma \vdash s \wedge t}{\Gamma \vdash s} \wedge^- \qquad \frac{\Gamma \vdash s \text{ Prop} \quad \Gamma, \neg s \vdash \bot}{\Gamma \vdash s} \neg^-$$

For many systems, the formation rules—which generate the well-formed expressions, including propositions—are *independent* of the rules governing other kinds of judgements. Such systems include those for which meaning-independent notion of syntax can be specified. But TPL also allows us to express formation rules that depend upon truth judgements, rather than on purely "syntactic" notions. We can exemplify this with a weak form of material implication, that only forms a proposition if the antecedent is true.

$$\frac{\Gamma \vdash s \text{ Prop} \quad \Gamma, s \text{ True} \vdash t \text{ Prop}}{\Gamma, s \text{ True} \vdash (s \to t) \text{ Prop}} \to^{\prime F}$$

TPL allows us to define various type systems. As an example, we can present a version of Simple Type Theory (Church, 1940), with entities $e$, propositions $p$, and functions $\langle S, T \rangle$ from $S$ to $T$.

$$\overline{\Gamma \vdash e \; \mathsf{Type}} \quad \overline{\Gamma \vdash p \; \mathsf{Type}}$$

$$\frac{\Gamma \vdash S \; \mathsf{Type} \quad \Gamma \vdash T \; \mathsf{Type}}{\Gamma \vdash \langle S, T \rangle \; \mathsf{Type}} \langle \cdot, \cdot \rangle^F \quad \frac{\Gamma \vdash f : \langle S, T \rangle \quad \Gamma \vdash a : S}{\Gamma \vdash fa : T} \circ$$

More complex types (and data structures) can be defined.

### 3.2 Questions and Answers in Typed Predicate Logic

We can represent questions using expressions of the form $[x : T \mid \phi]$. This is reminiscent of the notation of "schema" in computer science. The type $T$ indicates what type of term answer is appropriate. The proposition $\phi$ must be satisfied by any constituent answer that satisfies the specification. This can be a distinct representation for questions—no ontological reduction is required.[10]

$$\frac{\Gamma \vdash T \; \mathsf{Type}}{\mathsf{Quest}(T) \; \mathsf{Type}} \mathsf{Quest}(T)^F \quad \frac{\Gamma \vdash T \; \mathsf{Type} \quad \Gamma, x : T \vdash \phi \; \mathsf{Prop}}{\Gamma \vdash [x : T \mid \phi] : \mathsf{Quest}(T)} [\cdot \mid \cdot]^F$$

For answers, we adopt a form of structured proposition. We can represent structured propositions as $\langle f \mid t \rangle$, with a topic $t$ and focus $f$. This can be kept distinct from the notion of a pair.

$$\frac{\Gamma, x : T \vdash t(x) \; \mathsf{Prop} \quad \Gamma \vdash f : T}{\Gamma \vdash \langle f \mid t \rangle \; \mathsf{Prop}} \langle \cdot \mid \cdot \rangle^F$$

$$\frac{\Gamma, f : T \vdash t(f) \; \mathsf{Prop} \quad \Gamma \vdash a : T \quad \Gamma \vdash t(f)}{\Gamma \vdash \langle f \mid t \rangle} \langle \cdot \mid \cdot \rangle^+$$

$$\frac{\Gamma, f : T \vdash t(f) \; \mathsf{Prop} \quad \Gamma \vdash f : T \quad \Gamma \vdash \langle f \mid t \rangle}{\Gamma \vdash t(f)} \langle \cdot \mid \cdot \rangle^-$$

We can now introduce a relation, ans, that makes a well formed proposition between question and putative answer if they are of the appropriate nature. The judgement that $a$ is a *potential* answer to a question $q$

---

[10] Polymorphic typing can be used to account for the systematic behaviour of embedded interrogatives of distinct types (cf. Fox & Lappin, 2005, chapter 5).

is then captured by the well-formedness judgement of $a$ ans $q$—that is ($a$ ans $q$) Prop.

$$\frac{\Gamma, x : T \vdash t \text{ Prop} \quad \Gamma \vdash a : T}{\Gamma \vdash (\langle a \mid \lambda x.t \rangle \text{ ans } [x : T \mid t]) \text{ Prop}} \text{ ans}^F$$

This is the canonical propositional case. It would need to be generalised to include cases where the propositional part of the answer is "congruent" with the question, or is left unstated. If we have no analysis of ellipsis, this latter case could be approximated by allowing the topic in the answer to be optional.

If $a$ is a *potential* answer to a question $q$, then we can derive well-formedness judgement, ($a$ ans $q$) Prop. For it to be a *true* (or *correct*) answer, we need to be able to derive the judgement ($a$ ans $q$) True (cf. Karttunen, 1977).

$$\frac{\Gamma, x : T \vdash t \text{ Prop} \quad \Gamma \vdash a : T \quad \Gamma \vdash t[x/a]}{\Gamma \vdash (\langle a \mid [\lambda x.t] \rangle \text{ ans } [x : T \mid t]) \text{ True}} \text{ ans}^+$$

$$\frac{\Gamma, x : T \vdash t \text{ Prop} \quad \Gamma \vdash a : T \quad \Gamma \vdash (\langle a \mid [\lambda x.t] \rangle \text{ ans } [x : T \mid t]) \text{ True}}{\Gamma \vdash t[x/a]} \text{ ans}^-$$

It is worth observing that this framework also allows us capture fine-grained intentionality without possible worlds. Questions that have the same (possible) answers need not be equated. This is in part because we maintain an ontological distinction between questions and their (possible) answers.[11]

## 4 Summary

The claim being made here is not that it is possible or appropriate to dispense with all of the meta-theoretic narrative that accompanies any well-constructed formal semantic analysis. Rather, the argument is that those aspects of the narrative that seek to apply ontological classifications and distinctions ought to feature in the formal analysis, and that they can do so given an appropriate formalism. This avoids conflating the interpretations and patterns of behaviour of expressions that can result from reducing all formal meaning to the language of set theory and simple types.

In particular, the formal theory sketched here (section 3.2) illustrates that the notions of "being a question", and of "being an answer to a question",

---

[11]TPL similarly allows us to maintain fine-grained intensionality for propositions.

*can* be captured as first-class judgements *within* a formal analysis. It does not *prohibit* us from making ontological reductions if we choose. For example, (structured) propositions and questions could be given the same term representation, or the topic–focus structure could be expressed in terms of a type-membership judgement.

We can avoid conflating the interpretations and patterns of behaviour of expressions that can result from reducing all formal meaning to the language of set theory and simple types. If ontological reductions are to be made, it should through choice, not because they are imposed by the limitations of a particular semantic framework, or methodology.

## References

Aloni, M., & van Rooy, R. (2002). The dynamics of questions and focus. In B. Jackson (Ed.), *Proceedings of the twelfth Semantics and Linguistic Theory conference (SALT 12)*. Cornell University: CLC Publications.

Benacerraf, P. (1965). On what numbers could not be. *The Philosophical Review, 74*, 47–73.

Black, M. (1944). The paradox of analysis. *Mind, LIII*, 263–267.

Church, A. (1940). A formulation of the Simple Theory of Types. *The Journal of Symbolic Logic, 5*(2), 56–68.

Fox, C., & Lappin, S. (2005). *Formal foundations of intensional semantics*. Oxford: Blackwell.

Fox, C., & Turner, R. (2012). In defense of axiomatic semantics. In P. Stalmaszczyk (Ed.), *Philosophical and formal approaches to linguistic analysis* (pp. 145–160). Ontos Verlag.

Ginzburg, J. (1992). *Questions, queries and facts: A semantics and pragmatics for interrogatives*. Unpublished doctoral dissertation, Stanford University.

Ginzburg, J., & Sag, I. (2000). *Interrogative investigations: The form, meaning and use of English interrogatives*. Stanford: CSLI.

Groenendijk, J., & Stokhof, M. (1984). On the semantics of questions and the pragmatics of answers. In F. Landman & F. Veltman (Eds.), *Varieties of formal semantics* (pp. 143–170). Dordrecht: Foris.

Groenendijk, J., & Stokhof, M. (1997). Questions. In J. van Benthem & A. ter Meulen (Eds.), *Handbook of logic and linguistics* (pp. 1055–1124). Amsterdam: North Holland.

Halliday, M. A. K. (1967). Notes on transitivity and theme in English (part 2). *Journal of Linguistics*, *3*, 199–244.

Hamblin, C. L. (1958). Questions. *Australian Journal of Philosophy*, *36*, 159–168.

Hamblin, C. L. (1973). Questions in Montague English. *Foundations of Language*, *10*, 41–53.

Hausser, R. (1983). On questions. In F. Kiefer (Ed.), *Questions and answers* (pp. 97–158). Dordrecht: Reidel.

Hausser, R., & Zaefferer, D. (1979). Questions and answers in a context dependent Montague grammar. In F. Guenthner & M. Schmidt (Eds.), *Formal semantics and pragmatics for natural languages.* Dordrecht: Reidel.

Hull, R. (1975). A semantics for superficial and embedded questions in natural language. In E. Keenan (Ed.), *Formal semantics of natural language.* Cambridge University Press.

Jackendoff, R. S. (1972). *Semantic interpretation in generative grammar.* Cambridge, MA: MIT Press.

Karttunen, L. (1977). Syntax and semantics of questions. *Linguistics and Philosophy*, *1*, 3–44.

Krifka, M. (1991). A compositional semantics for multiple focus constructions. In S. K. Moore & A. Z. Wyner (Eds.), *Proceedings of the first Semantics and Linguistic Theory conference (SALT I)* (pp. 17–53).

Krifka, M. (2001). For a structured meaning account of questions and answers. In C. Fery & W. Sternefeld (Eds.), *Audiatur Vox Sapientia. A Festschrift for Arnim von Stechow* (pp. 287–319). Berlin: Akademie Verlag.

Paul, H. (1880). *Prinzipien der Sprachgeschichte.* Tübingen: Max Niemeyer. (English translation from the second edition (1890) by H. A. Strong: H. Paul (1970). *Principles of the History of Language.* College Park, Maryland: McGrath Publishing Company.)

Roberts, C. (1996). Information structure in discourse: towards an integrated formal theory of pragmatics. In J. H. Yoon & A. Kathol (Eds.), *Ohio working papers in semantics* (Vol. 49). Columbus: Ohio State University.

Rooth, M. (1992). A theory of focus interpretation. *Natural Language Semantics*, *1*, 75–116.

Stalnaker, R. C. (1978). Assertion. In P. Cole (Ed.), *Syntax and semantics* (Vol. 9, pp. 315–332). New York: AP.

Tichý, P. (1978). Questions, answers, and logic. *American Philosophical*

*Quarterly, 15*(4), 275–284.

Turner, R. (2008). Computable models. *Journal of Logic and Computation, 18*(2), 283–318.

Turner, R. (2009). *Computable models.* London: Springer-Verlag.

Vallduví, E. (1992). *The informational component.* New York: Garland Press.

Vallduví, E. (1993). *Information packaging: A survey* (Tech. Rep.). University of Edinburgh. (A report prepared for WOPIS — Word Order, Prosody, and Information Structure.)

von Stechow, A. (1981). Topic, focus, and local relevance. In W. Klein & W. Levelt (Eds.), *Crossing the boundaries in linguistics.* Dordrecht: Reidel.

von Stechow, A. (1982). *Structured propositions.* (Manuscript)

von Stechow, A. (1991). Focusing and backgrounding operators. In W. Abraham (Ed.), *Discourse particles* (pp. 37–84). Amsterdam: John Benjamins.

von Stechow, A., & Zimmermann, T. E. (1984). Term answers and contextual change. *Linguistics, 22,* 3–40.

Chris Fox
CSEE, University of Essex
Wivenhoe Park, Colchester, CO4 3SQ, UK
e-mail: foxcj@essex.ac.uk
URL: http://chris.foxearth.org

# Are Impossible Worlds Trivial?

MARK JAGO

**Abstract:** Theories of content are at the centre of philosophical semantics. The most successful general theory of content takes contents to be sets of possible worlds. But such contents are very coarse-grained, for they cannot distinguish between logically equivalent contents. They draw intensional but not *hyperintensional* distinctions. This is often remedied by including *impossible* as well as possible worlds in the theory of content. Yet it is often claimed that impossible worlds are metaphysically obscure; and it is sometimes claimed that their use results in a trivial theory of content. In this paper, I set out the need for impossible worlds in a theory of content; I briefly sketch a metaphysical account of their nature; I argue that worlds in general must be very fine-grained entities; and, finally, I argue that the resulting conception of impossible worlds is not a trivial one.

**Keywords:** impossible worlds, content, hyperintensionality, semantics

## 1 Introduction

Theories of content are at the centre of philosophical semantics. One task is to assign particular contents to particular kinds of expressions. But a prior, more basic task is to give a theory of what contents in general are. This is a metaphysical endeavour, in that the question concerns the *nature* of contents, although it must answer to the semantic data. The most successful and thoroughgoing theory of content currently on offer treats the contents of sentences as sets of possible worlds (Lewis, 1986; Stalnaker, 1976a, 1976b). These contents can be thought of as *propositions*. Contents of sub-sentential terms can then be thought of as functions on possible worlds which, when combined in a way that mirrors the syntax of the corresponding sentence, produces the content (a set of worlds) of that sentence (von Fintel & Heim, 2007).

This approach to content meshes very well with the *modal epistemic logic* approach to modelling knowledge and belief (Hintikka, 1962), in terms of epistemic and doxastic accessibility relations between worlds. Indeed, if $R_i$ is agent $i$'s epistemic accessibility relation then, given a world $w$, the set of worlds $\{u \mid Rwu\}$ can be thought of as the content of $i$'s epistemic state at $w$. She knows that $A$ at $w$ iff the proposition $\langle A \rangle$ includes that content, i.e. $\{u \mid Rwu\} \subseteq \langle A \rangle$. The possible worlds theory of content also meshes very well with Bar-Hillel and Carnap's (1953) analysis of information.

The major thorn in the possible worlds analysis of content (and of knowledge, belief, information and cognate notions) is that it is able to analyse intensional but not *hyperintensional* operators. An operator '$O$' is hyperintensional iff '$OA$' and '$OB$' can take different values for some logically equivalent '$A$' and '$B$'. 'Knows that', 'believes that', 'has the information that' and 'has the content that' are all hyperintensional operators (although many theorists have tried to argue to the contrary). In the case of knowledge, for example, the non-hyperintensionality of the possible worlds account results in the notorious *logical omniscience* problem (Hintikka, 1975; Stalnaker, 1991), whereby agents are treated as automatically knowing all logical truths and all consequences of what they know.

One solution to these issues (hyperintensionality in general and logical omniscience in particular) is to supplement the domain of possible worlds with *impossible* worlds. These are worlds according to which impossible things happen. They stand to Escher drawings as possible worlds stand to depictions of possible situations. In my view, the addition of impossible worlds to a worlds-based approach provides the best account we can give of epistemic and doxastic notions of content, including knowledge and belief states (Hintikka, 1975), cognitive significance and information (Chalmers, 2010; Jago, 2009a), and the content of informative deduction (Jago, 2012c).

Yet acceptance of impossible worlds in philosophical theorising is far from the norm. We lack a persuasive metaphysical theory of their nature and of how they represent. There are, moreover, worries that impossible worlds either fail to overcome the problems they were designed to solve or else are too trivial to deserve our respect.

My aims in this paper are to set out the need for impossible worlds in a theory of content (§2); to sketch briefly a metaphysical account of the nature of possible and impossible worlds (§3); to argue for a highly fine-grained theory of impossible worlds (§4) and, finally, to argue that the resulting conception of impossible worlds is not a trivial one (§5).

## 2 The Need for Impossible Worlds

The case for impossible worlds (in addition to possible worlds) can be made by (i) highlighting deficiencies in the possible worlds account, and (ii) arguing that the addition of impossible worlds best overcomes these problems. The problems I have in mind concern hyperintensional concepts such as knowledge, belief, information and content in general. In this section, I'll review those problems and argue briefly that the impossible worlds approach is the best solution.

*Knowledge:* Suppose you and I play a game of chess (with no time controls) in which a draw counts as a win for black. It is then a surprising mathematical fact that, at any stage of the game, one of us has a *winning strategy*: there is a function from the game's previous moves to that player's next move which guarantees victory, regardless of how the other player plays. Suppose $\sigma$ is a winning strategy for me, which tells me to move my queen to e7 next. Yet I don't move my queen to e7 and go on to lose the game. Had I known that Qe7 was recommended by a winning strategy, I would have made that move. So it is clear that I *didn't* know that $\sigma$ is a winning strategy, even though $\sigma$'s being a winning strategy for me followed mathematically from the state of the game at that time.[1]

*Information:* Gaining information amounts to narrowing down one's set of epistemically accessible worlds. But no logical or mathematical truth rules out any (logically) possible world. So the possible worlds approach must say, with Wittgenstein (1921/1922, §6.11, §2.19, §6.12), that all logical and mathematical truths are utterly uninformative (and, according to Wittgenstein, unsurprising too). But of course there *are* surprises in logic and mathematics, as a cursory leaf through the technical literature shows.[2] Students who happily accept the truth-table for the material conditional '$\to$' are surprised to learn that, for any sentences '$A$' and '$B$' whatsoever, in any situation either '$A \to B$' or '$B \to A$' will be true. Even preeminent logicians are occasionally surprised by their results, as the reaction to early

---

[1] If I really did know all consequences of what I know, I will know a winning strategy in 50% of these chess games. But then I would know how to win at least 50% of the games, *regardless of who my opponent is!* It seems to me outlandish folly of the highest order to think that I could *ever* beat Kasparov or Fischer, let alone in 50% of these games!

[2] As Dummett says, "when we contemplate the wealth and complexity of number-theoretic theorems, ... we are struck by the difficulty of establishing them and *the surprises they yield*" (Dummett, 1978, p. 297, my emphasis).

results in model theory shows.[3] A natural way to explain what makes a given truth surprising is to appeal to its informativeness. If logical results were not informative, then they would never be surprising. So logical truths can be informative.

*Content:* The possible worlds approach treats propositions as sets of possible worlds. As a consequence, logically equivalent propositions are numerically identical to one another. This is Stalnaker's principle (I) (1976a, p. 9). But the principle is incorrect. We can bring out what's wrong with it by focusing on what a given proposition (or the sentence expressing it) is *about*, or by focusing on what *makes* a given proposition true. One job of a proposition is to give the content of the sentence expressing it, and in particular, to specify what that sentence is about. Hence, if sentences '$A$' and '$B$' express the same proposition in a common context $c$, then they should be about the very same things (in $c$).

As a consequence, the possible worlds approach entails that logically equivalent sentences are always about the very same things (in any context). But this is not so. The sentence

(1) Puss is crafty $\lor$ Puss isn't crafty

is about Puss (and perhaps about *craftiness* too), but not Rover. By contrast,

(2) Rover is snoring $\lor$ Rover isn't snoring

is about Rover (and perhaps about *snoring* too), but not Puss. Thus (1) and (2) are about completely different things. Yet (1) and (2) are logically equivalent, contradicting the consequence of the possible worlds approach.

We can make essentially the same point by focusing on the notion of what makes a proposition true or false. The fact *that Puss is crafty* (but not the fact *that Rover is snoring*) makes (1) true. The fact *that Rover is snoring* (but not the fact *that Puss is crafty*) makes (2) true. Hence by Leibniz's Law (1) and (2) express distinct propositions, despite being logically equivalent. So propositions cannot be sets of possible worlds.

*Counterfactuals:* A worlds-based account provides the best semantics for counterfactuals (Lewis, 1973; Stalnaker, 1968). Yet to make good sense of *counter-possible* conditionals, such as

---

[3]Löwenheim's theorem perplexed Skolem, who gave the first correct proof of it (Skolem, 1922). Skolem took the result to be paradoxical.

(3) If Linear Logic had been the One True Logic, then the TONK rules would have been valid rules of inference.

(4) If Fermat's Last Theorem had been false, then I would have been a lemon.

we require impossible as well as possible worlds to be part of the story.[4] Both (3) and (4) are trivially true, according to the Stalnaker-Lewis approach; yet each is clearly false. On the worlds-based approach, false counterfactuals require worlds where the antecedent is true and the consequent is false. Such worlds are, of course, impossible; hence the need for impossible worlds.

One can of course resist these moves in various ways. One could flatly deny that epistemic and counterfactual concepts are hyperintensional.[5] (I find that view implausible.) Or one could adopt a *structuralist* approach, by supplementing the possible worlds account with linguistic structure, rather than impossible worlds.[6] Or one could abandon worlds-based approaches altogether.

The structuralist approach on its own does not make all of the fine-grained distinctions between contents that we require. Consider:[7]

(5) All woodchucks are woodchucks.

(6) All woodchucks are whistle-pigs.

Since whistle-pigs are woodchucks (and necessarily so), a structuralist view which takes semantic values (written '$[\![-]\!]$') of nouns to be extensions or intensions will hold that $[\![\text{woodchucks}]\!] = [\![\text{whistle-pigs}]\!]$. If so, both (5) and (6) express the proposition:[8]

---

[4]Brogaard and Salerno (2008), Nolan (1997), Read (1995) and Routley (1989) discuss the impossible-worlds approach to counter-possible conditionals.

[5]Lewis (1996) and Stalnaker (1984) defend this view. Stalnaker in particular is at pains to explain away the *appearance* of hyperintensionality via a metalinguistic approach. But my failure to know how best to proceed in chess is *not* a failure to grasp what words mean.

[6]Chalmers (2011), Cresswell (1985), King (2007), Salmon (1986) and Soames (1987) defend structuralist theories.

[7]The example is Dave Ripley's (2012).

[8]I'm assuming (with Ripley, 2012) that the relevant syntactic structure here consists of a quantifier-phrase 'All woodchucks' and a complement 'are woodchucks', rather than a universal quantifier '$\forall x$' attached to an open sentence '$x$ is a woodchuck'.

(5′) ⟨⟨⟦All⟧, ⟦woodchucks⟧⟩, ⟨⟦are⟧, ⟦woodchucks⟧⟩⟩

Yet, whilst everyone knows (5), one might not believe (6). Ripley (2012, p. 9) considers the case of Tama, who 'knows he is allergic to whistle-pigs, and knows that he has just been bitten by a woodchuck'. Tama fears that (6) is true, but surely no one fears that (6), as a trivial logical truth, is true! So the structuralist approach is not suitably hyperintensional for our needs.

Contrast this with the impossible worlds approach. It is impossible that whistle-pigs are not woodchucks; so, according to some impossible world, whistle-pigs are not woodchucks. There are also incomplete worlds, representing nothing about woodchucks. Hence there are worlds enough to provide distinct contents for (5) and (6).

The final option, of abandoning the worlds-based approach altogether, is extreme. Hyperintensionality worries aside, the approach is the most comprehensive and systematic account of content on offer. It is one of the key theories in formal semantics and is used widely in computer science, artificial intelligence and game theory. We should not abandon the benefits that the theory brings lightly. As always in science, the rational approach is to begin with our best theory and modify it so as to include phenomena which it does not currently accommodate. For the remainder of the paper, therefore, I will assume that there is desirable theoretical utility to be gained from working with impossible as well as possible worlds.

## 3 The Nature of Worlds

One might take non-actual worlds to exist in much the same way that our own world—the universe in its entirety—exists. I'll call such worlds *genuine* worlds. Lewis (1973, 1986) is the main proponent of this view in the case of possible worlds; Yagisawa (1988, 2010) is its champion in the case of impossible (as well as possible) worlds.[9] An opposing *actualist* view holds that the non-actual worlds are mere *representations*, or models, of the ways our universe could or could not be.[10] There are many of varieties of ersatz world on offer to the theorist, depending on how she wants to represent ways the universe could or could not be.[11]

---

[9] See (Jago, 2012a) for the case against Yagisawa's account of impossible worlds.

[10] Genuine worlds are also representations of ways things could or could not be, but they are not *mere* representations. They do not represent in a pictorial or linguistic way, as ersatz words do.

[11] A third alternative, taking a cue from Bolzano (1834) and Meinong (1904), is to accept that there are worlds other than our own but deny that such worlds *exist* (Priest, 2005). But

What marks a world as genuine or ersatz is how that world *represents*. A genuine world represents the existence of a flying hippo by having a flying hippo as a part.[12] Ersatz worlds, by contrast, represent the existence of a flying hippo by picturing, or linguistically describing, a flying hippo. As a consequence, Lewisian genuine worlds obey the *exportation principle*: if world $w$ represents something as being an $F$, then something is an $F$. For if a genuine world $w$ represents something as being $F$, then $w$ contains an $F$ as a part. And as $w$ is part of the totality of being, that particular $F$ too is part of the totality of being: so something is an $F$.

The exportation principle is problematic for any account of impossible genuine worlds, as Lewis (1986) notes. Exporting merely possible entities (or states of affairs) from genuine possible worlds lumbers us with a large and counterintuitive but still consistent ontology. Exporting impossible entities (or states of affairs) from genuine impossible worlds, by contrast, drags us into contradiction. If there is an impossible genuine world according to which there is a round square, then (given exportation) there is a real entity which is both round and square. But it is a necessary truth that no square is round, and so our exported round entity is also not round: contradiction![13] So one cannot accept impossible genuine worlds on the Lewisian model.

To avoid the exportation worry, must deny the move from 'according to $w$, $Ax$' to 'something is such that $Ax$'. One way to do this (for genuine worlds) is to insist that all property possession is world-relative: $x$ may be $F$-at-$w$ but not $F$-at-$w'$ (much as I am happy on Sunday but not on Monday), with nothing being $F$ *simpliciter*. McDaniel (2004) defends a view along these lines. These *overlapping* genuine worlds are just as unsuitable for providing impossible worlds as Lewis's worlds, although for a different reason. Take the case of Richard Sylvan, the New Zealand logician, born Richard Routley. It is impossible for Sylvan to be other then Routley, and hence there is an impossible world $w$ at which Sylvan is not identical to Routley. On the view under consideration, that is to say that Sylvan bears the *non-identical-to-Routley* relation to world $w$. But Routley does not bear the *non-identical-to-Routley* relation to world $w$ and hence, by Leibniz's

---

there remains the question of the nature of such worlds. If they are genuine worlds, and genuine worlds are the best theoretical tool for the job, then I'd prefer to say that such entities exist.

[12] More precisely, genuine worlds represent *de dicto* that $A$ by being such that $A$. But they may represent *de re* of $x$ that $Ax$ by being such that, of some counterpart $y$ of $x$, $Ay$ (Lewis, 1968, 1986).

[13] The problem also affects the Bolzano-Meinong approach to impossible worlds. If such worlds are genuine (albeit non-existent) parts of reality, we can infer that reality itself has contradictory parts, described truly by some contradiction.

law, Sylvan and Routley are not identical.[14] But this is absurd: Routley *is* (or *was*) Sylvan! Consequently, genuine worlds (whether overlapping or Lewisian) are not a suitable treatment of impossible worlds.

Ersatz worlds have a clear advantage over genuine worlds when it comes to impossibilities, for they do not require anything to *be* impossible, or possess impossible properties (either *simpliciter* or at-a-world). They are *mere* representations, and mere representations of impossibility are commonplace. (Just look at an Escher drawing, or read the Bible.) Given that we want very fine-grained impossible worlds (to make sense of the content of epistemic and doxastic states), the linguistic approach is the way to go. Impossibilities may be incomplete as well as inconsistent, and linguistic representation accounts for incompleteness far better than pictorial representation does. So I will focus on the linguistic approach to constructing ersatz worlds.[15]

On the linguistic approach, ersatz worlds are sets of sentences in some 'worldmaking' language. It must be clear how this language is to be interpreted. The simplest approach is to use a *Lagadonian* language, in which we take particulars to be names and properties and relations to be predicates, each interpreted to refer to itself (Carnap, 1947; Lewis, 1986). Atomic sentences are sequences of an $n$-place predicate (that is, a property or relation) followed by $n$ terms (particulars). Designated set-theoretic constructions will serve as connectives, quantifiers and variables. We allow quantifier-prefixes to be infinitely long and conjunction and disjunction symbols to operate on (possibly infinite) sets of sentences. In this way, we avoid the *cardinality objection* from Lewis (1973, p. 90) and Bricker (1987, pp. 340–343).[16]

This language must represent distinct (possible or impossible) situations without conflation. This requirement is much harder to meet, for there is

---

[14] More formally: there is an impossible world $w$ such that $s$ bears $\lambda xx \neq r$ to $w$ but $r$ does not bear $\lambda xx \neq r$ to $w$:

$$[\lambda y R(y, \lambda xx \neq r, w)]s \wedge \neg[\lambda y R(y, \lambda xx \neq r, w)]r$$

which entails $r \neq s$.

[15] Linguistic ersatz proposals were put forward, in different ways and for different purposes, by Carnap (1947) and Hintikka (1962, 1969). The approach has been defended more recently by Melia (2001) and Sider (2002).

[16] The objection, in short, is that there are more possibilities (at least $\beth_2$) than sets of sentences (at most $\beth_1$), and hence sets of sentences cannot represent *all* the possibilities without conflation. But the objection assumes a countable worldmaking language, and so it is ineffective against the above proposal.

'an apparently devastating problem', the 'problem of descriptive power', for linguistic ersatzism Sider (2002, p. 281). The problem is to represent possible but non-actual particulars, properties and relations *without conflation*. Since (by actualist assumptions) such entities do not exist, they cannot be invoked as names for themselves. So, it seems, we cannot represent such mere possibilities by *naming* them. The problem is indeed hard, and there is no space to assess it properly here; I set out my solution in detail elsewhere (Jago, 2012b).

## 4 The Granularity of Worlds

Worlds are sets of sentence (of the worldmaking language). But *which* sets of sentences count as worlds? For Cresswell (1973), impossible worlds are worlds governed by some non-classical (e.g., intuitionistic or paraconsistent) logic. For other authors, they are worlds which represent some contradiction '$A \wedge \neg A$' as being true (Berto, 2010; Lycan, 1994). Combining these ideas, it is natural to think of impossible worlds as corresponding to the relational models of paraconsistent logic (see, e.g., Priest, 1987). On this view, worlds are maximal and closed under paraconsistent consequence: if $\Gamma \subseteq w$ and $\Gamma$ paraconsistently entails '$A$', then '$A$' $\in w$ too.

This is not a good move. Paraconsistent logic invalidates *modus ponens* and *disjunctive syllogism*. So the present approach will treat those inferences as being potentially informative. Yet the standard conjunction rules and *disjunction introduction* remain paraconsistently valid and hence can never be informative, on the present approach. Why think that *modus ponens* is informative, if *conjunction elimination* is not? After all, *modus ponens* is just as integral to the meaning of '$\rightarrow$' as *conjunction elimination* is to the meaning of '$\wedge$', and so there is just no reason for declaring the former but not the latter to be informative.[17]

There is another, deeper problem with the selected worlds approach: it fails to explain why the worlds it provides are suitable tools for analysing epistemic notions. An epistemically accessible world is one which *looks* possible to the agent in question. But a world that is *trivially* impossible won't look possible to a minimally rational agent. On the present approach,

---

[17]Note that the proponent of the selected worlds approach cannot respond by claiming that the meaning of '$\rightarrow$' is such that *modus ponens* is invalid. That would certainly provide a reason for not treating *modus ponens* and *conjunction elimination* on an informational par, but it would just as clearly hinder the task of explaining how *valid* inferences can be informative.

worlds may represent explicitly contradictory situations $A \wedge \neg A$. Such situations are trivially impossible, and hence such worlds should never be considered epistemically possible, for any rational agent. If we remove such worlds from the account, however, we are left only with classical possible worlds, and the problems of logical omniscience return.

The approach I favour treats *any* set of worldmaking sentences as a world (Priest, 2005), such that $w$ represents that $A$ iff there is a worldmaking sentence $S \in w$ which represents that $A$. Consequently, many worlds are *incomplete*, in the sense that (for some '$A$') they represent neither that $A$ nor that $\neg A$. A world is *prime* iff it contains a disjunct of each disjunction it contains. A world is *maximal* iff it contains either '$A$' or its negation, for each sentence '$A$' (of the worldmaking language). A world is *logically possible* (with respect to some logic) iff it is both maximal and consistent (with respect to that logic's derivability relation). Just when a world represents a *genuine metaphysical possibility* is a thorny issue, and one I do not propose to say anything about here. Ersatzers typically resort to primitive modal facts.[18] That's fine with me: my concern is not to give a reductive account of modal talk.

We can formulate a general argument in favour of this very liberal view of worlds. Suppose that knowledge is closed under the standard introduction and elimination rules for some connective $C$. This could be the case in general only because of the meaning of '$C$' and the kind of mental state that knowledge is. Those rules stand to the meaning of $C$ just as the standard introduction and elimination rules for other connectives $C'$ stand to the meaning of $C'$. Hence knowledge must be closed under the standard introduction and elimination rules for $C'$ too, for any standard connective $C'$. But these rules, taken together, are deductively complete and hence knowledge must be deductively closed. But knowledge is not deductively closed. By *reductio*, knowledge is not closed under the standard introduction and elimination rules for any (non-trivial) connective. Notice that this argument does not assume a worlds-based account of knowledge. It provides an independent reason for thinking it possible for an agent to know that $A$ but not that $B$, for any distinct '$A$' and '$B$'. To capture this feature in a worlds-based account, we require worlds which are not closed under any inference rule (except *identity*, $A \vdash A$). In other words, we must count any set of worldmaking sentences as a world.

---

[18]But, as Sider (2002, p. 306) says, 'a reduction of talk of possibilia that employs primitive possibility and necessity is nevertheless valuable since talk of possibilia runs beyond what can be said in the language of quantified modal logic'.

This very fine-grained notion of worlds (which, in my view, is mandatory for a suitable worlds-based account of content) does, however, invite worries about triviality. If *any* set of sentences counts as a world, then how can sets of worlds tell us anything about the *content* of a sentence? That, in essence, is the worry I address in the next section.

## 5  Is the Approach Trivial?

I will end by considering a general objection to the kind of approach I favour, but which might be raised against any semantic account which uses very fine-grained impossible worlds. The objection goes roughly as follows:[19]

> We want to assign a content to a sentence '$A$'. We first try assigning a set of possible worlds, but we soon see that this is too coarse-grained. So we finesse the approach and assign a set of possible and impossible worlds to '$A$'. These worlds are sets of sentences, and a world $w$ is in the content of '$A$' if and only if that world is a set containing '$A$'. But then we come full circle: the 'content' assigned to '$A$' is none other than '$A$' itself, with a bit of set-theoretic machinery thrown in for good measure. Clearly, this 'content' can tell us nothing about the *meaning* of '$A$'. We asked a question about the meaning of '$A$' and what we get back by way of answer, more or less, is '$A$' itself!

There are really two worries expressed here, one to do with defining content in terms of linguistic entities, and one to do with the granularity of contents thus assigned. These distinct worries might be expressed as follows.

WORRY 1: A sentence '$A$' is assigned a content, which is a set of worlds, which themselves are sets of sentences. So ultimately, the content of '$A$' is given in terms of further sentences, much like a translation of '$A$' into some other language. But a translation of '$A$' can tell us what '$A$' means (or what its content is) only if we have a prior grasp of the meanings (or content) of the sentences used to translate '$A$'. Ultimately, we have to step beyond the linguistic realm of translations

---

[19] I have heard objections of this form several times in discussion, although I have not seen it in print. It seems to me to be a genuine problem for *some* hyperintensional theories and so it is worth addressing here.

and assign meanings and contents *non-linguistically*, by correlating the sentence with the non-linguistic world.

WORRY 2: A sentence '$A$' is assigned a content, which is the set of worlds which represent that $A$. But these worlds are so fine-grained that the content thus assigned does not even include the content of '$A \vee B$', or of other trivial consequences of '$A$'. So the *granularity* of contents is just the granularity of sentences. Each sentence of the language has its own unique content. So why not just let the content of '$A$' be the singleton $\{`A`\}$? Of course, if we did that, the theory would unquestionably *not* be an adequate account of content. But how is the present proposal any better?

Both worries, although rather vague, need to be taken seriously. A weak(ish) response to WORRY 1 goes as follows:

> We are interested in *modelling* the contents of sentences (and of epistemic and doxastic states, and of thoughts in general). We do this in terms of worlds. It really doesn't matter what those worlds *are*; all that matters is their formal properties. As in modal logic, it doesn't matter what we take the worlds to be. All that matters is the logics obtained by imposing various constraints on the semantics, and we can investigate all of this without being concerned one jot with the metaphysical nature of worlds.

It is indeed often a good idea to keep logical and metaphysical issues separate when investigating modal, intensional and hyperintensional concepts. But at some point, those with curious minds want to know, what *is* content? It is part of an answer to say that contents have such-and-such structure, but this is only part of the answer. Similarly, it is one thing to know that the natural numbers have an $\omega$-structure, but that on its own does not tell us what numbers *are*.

What WORRY 1 misses, however, is that the account of content I suggested above *does* make contact with the non-linguistic world. Worlds are constructed from sets of worldmaking sentences, but those sentences are themselves constructed from worldly entities, in the Lagadonian way. The worldmaking sentence representing that Bertie is adorable contains Bertie himself and the property of *being adorable*. English sentences containing

'Bertie' are assigned a content which contains Bertie himself, in all his panting, muddy, tail-wagging glory (just as they are on the Russellian view of propositions).

WORRY 2 is more subtle. If the objection is that the account I've given assigns contents which are *too fine-grained*, then we need to hear more from the objector about what is the correct granularity to assign to contents. Suppose, for example, she says it is the coarse granularity provided by metaphysical or logical equivalence, or the granularity provided by the paraconsistent approach discussed in §4. But I argued above (§4) that such notions of content are always too coarse grained and so I take it that WORRY 2 should not be put in these terms.

Perhaps the genuine worry contained in WORRY 2 is that, on my view, a sentence '$A$' is assigned a content in terms of *itself*, or in terms of the singleton $\{`A'\}$, with some set-theoretic distractions thrown in. But this is clearly a misinterpretation of the proposal. '$A$' (a sentence of the object language, English, say) is not assigned a content in terms of '$A$' or $\{`A'\}$, but rather in terms of (sets of) Lagadonian *worldmaking* sentences. Hence it is not true that each *object language* sentence is assigned a content which ultimately consists of itself, plus some set-theoretic smoke and mirrors.[20]

The possible-and-impossible worlds account of content is not trivial. It succeeds in linking sentences to non-linguistic reality, as any account of content should. It provides a clear and simple account of when a content is true at a give world: content $C$ is true according to world $w$ iff $w \in C$.

One issue I have not touched on here is that, since every set of worldmaking sentences constitutes a world, there exist worlds which include explicit contradictions, '$A \wedge \neg A$'. Such worlds should not be considered epistemically accessible for any possible, minimally rational agent (§4). Nevertheless, as I argued in §4, such sets of worldmaking sentences need to be counted as worlds if we are to provide suitable contents for contradictory natural language sentences. They are worlds that are never epistemically accessible. The problem is to draw a distinction between trivially and non-trivially impossible worlds, and to allow only the latter to be epistemically accessible. This is a difficult task, to be sure. I discuss it in detail and present a solution elsewhere (Jago, 2009a, 2009b, 2012c).

---

[20]Except in the case where the object language itself is the worldmaking language, of course. But why should we worry about that, given that the worldmaking language is fully interpreted in the Lagadonian way by letting each actual entity refer to itself?

## References

Bar-Hillel, Y., & Carnap, R. (1953). An outline of a theory of semantic information. In Y. Bar-Hillel (Ed.), *Language and information: Selected essays on their theory and application* (pp. 221–274). London: Addison-Wesley.

Berto, F. (2010). Impossible worlds and propositions: Against the parity thesis. *The Philosophical Quarterly, 60*(240), 471–486.

Bolzano, B. (1834). *Lehrbuch der Religionswissenschaft: Ein Abdruck der Vorlesungshefte eines ehemaligen Religionslehrers an einer katholischen Universität*. J.E.v. Seidel.

Bricker, P. (1987). Reducing possible worlds to language. *Philosophical Studies, 52*(3), 331–355.

Brogaard, B., & Salerno, J. (2008). Remarks on counterpossibles. In J. van Bentham, V. Hendricks, J. Symons, & S. A. Pedersen (Eds.), *Between logic and intuition: David Lewis and the future of formal methods in philosophy*. Synthese Library.

Carnap, R. (1947). *Meaning and necessity*. University of Chicago Press.

Chalmers, D. (2010). The nature of epistemic space. In A. Egan & B. Weatherson (Eds.), *Epistemic modality*. Oxford University Press.

Chalmers, D. (2011). Propositions and attitude ascriptions: A Fregean account. *Noûs, 45*(4), 595–639.

Cresswell, M. (1973). *Logics and languages*. London: Methuen.

Cresswell, M. (1985). *Structured meanings*. Cambridge, MA: MIT Press.

Dummett, M. (1978). The justification of deduction. In *Truth and other enigmas* (pp. 166–185). Cambridge, MA: Harvard University Press.

Hintikka, J. (1962). *Knowledge and belief: An introduction to the logic of the two notions*. Ithaca, N.Y.: Cornell University Press.

Hintikka, J. (1969). *Models for modalities: Selected essays*. D. Reidel Dordrecht.

Hintikka, J. (1975). Impossible possible worlds vindicated. *Journal of Philosophical Logic, 4*, 475–484.

Jago, M. (2009a). Logical information and epistemic space. *Synthese, 167*(2), 327–341.

Jago, M. (2009b). Resources in epistemic logic. In J.-Y. Béziau & A. Costa-Leite (Eds.), *Dimensions of logical concepts* (Vol. 55, pp. 11–33). Campinas, Brazil: Coleção CLE.

Jago, M. (2012a). Against Yagisawa's modal realism. *Analysis, 73*(1),

10–17.
Jago, M. (2012b). Constructing worlds. *Synthese*, *189*(1), 59–74.
Jago, M. (2012c). The content of deduction. *Journal of Philosophical Logic*, (published online, February).
King, J. (2007). *The nature and structure of content*. Oxford University Press.
Lewis, D. (1968). Counterpart theory and quantified modal logic. *The Journal of Philosophy*, *65*(5), 113–126.
Lewis, D. (1973). *Counterfactuals*. Harvard University Press.
Lewis, D. (1986). *On the plurality of worlds*. Oxford: Blackwell.
Lewis, D. (1996). Elusive knowledge. *Australasian Journal of Philosophy*, *74*(4), 549–567.
Lycan, W. (1994). *Modality and meaning*. Dordrecht: Kluwer.
McDaniel, K. (2004). Modal realism with overlap. *Australasian Journal of Philosophy*, *82*(1), 137–152.
Meinong, A. (1904). Über Gegenstandstheorie. In A. Meinong (Ed.), *Untersuchungen zur Gegenstadstheorie und Psychologie*. Leipzig: Barth.
Melia, J. (2001). Reducing possibilities to language. *Analysis*, *61*(1), 19–29.
Nolan, D. (1997). Impossible worlds: A modest approach. *Notre Dame Journal of Formal Logic*, *38*(4), 535–572.
Priest, G. (1987). *In contradiction: A study of the transconsistent*. Dordrecht: Martinus Nijhoff.
Priest, G. (2005). *Towards non-being*. Oxford: Clarendon Press.
Read, S. (1995). *Thinking about logic*. Oxford: Oxford University Press.
Ripley, D. (2012). Structures and circumstances: Two ways to fine-grain propositions. *Synthese special issue on Propositions and Same-Saying*, *189*(1), 97–118.
Routley, R. (1989). Philosophical and linguistic inroads: Multiply intensional relevant logics. In J. Norman & R. Routley (Eds.), *Directions in relevant logic*. Dordrecht: Kluwer.
Salmon, N. (1986). *Frege's puzzle*. MIT press/Bradford books.
Sider, T. (2002). The ersatz pluriverse. *The Journal of Philosophy*, *99*(6), 279–315.
Skolem, T. (1922). Einige Bemerkungen zur axiomatischen Begründung der Mengenlehre. In *Selected works in logic* (pp. 137–152). Oslo: Universitetsforlaget.
Soames, S. (1987). Direct reference, propositional attitudes and semantic content. *Philosophical Topics*, *15*, 47–87.

Stalnaker, R. (1968). A theory of conditionals. In N. Rescher (Ed.), *Studies in logical theory* (pp. 98–112). Oxford: Blackwell.
Stalnaker, R. (1976a). Possible worlds. *Noûs*, *10*(1), 65–75.
Stalnaker, R. (1976b). Propositions. In A. MacKay & D. Merrill (Eds.), *Issues in the philosophy of language* (pp. 79–91). New Haven: Yale University Press.
Stalnaker, R. (1984). *Inquiry*. Cambridge, MA: MIT Press.
Stalnaker, R. (1991). The problem of logical omniscience I. *Synthese*, *89*, 425–440.
von Fintel, K., & Heim, I. (2007). *Intensional semantics*. (Available at http://tinyurl.com/intensional)
Wittgenstein, L. (1921/1922). *Tractatus logico-philosophicus*. Routledge & Kegan Paul.
Yagisawa, T. (1988). Beyond possible worlds. *Philosophical Studies*, *53*(2), 175–204.
Yagisawa, T. (2010). *Worlds and individuals, possible and otherwise*. Oxford University Press.

Mark Jago
Department of Philosophy
University of Nottingham
University Park
Nottingham NG7 2RD, UK
e-mail: mark.jago@nottingham.ac.uk
URL: www.markjago.net

# Alleged(ly)

## BJØRN JESPERSEN[1]

**Abstract:** Modal modifiers oscillate between being subsective and being privative. I offer an analysis of the property modifier *alleged*, as referred to in "*a* is an alleged assassin", and the propositional modifier *allegedly*, as referred to in "Allegedly, *a* is an assassin". I show, contrary to the received view, that propositions involving modal modification do yield non-trivial entailments that uniquely characterize them. This paper fills a logical gap in the model-theoretic edifice of modification developed by Montague, Kamp, Partee and others.

**Keywords:** modal modifiers, *alleged*, *allegedly*, Partee

## 1 Introduction

Modal modifiers such as *alleged* and *maybe* are the odd ones out among property and propositional modifiers. One fundamental logical fact about modal modifiers is that if $a$ is an alleged assassin then it can neither be ruled in nor ruled out that $a$ is an assassin. This feature renders them barely informative in terms of deriving deductively valid conclusions. In fact, modal modifiers are the only modifiers to be non-monotonic. Partee claims, not unreasonably:

> There is no meaning postulate for the *modal* adjectives, since they have no entailments — an *alleged murderer* may or may not be a murderer, and similarly for adjectives like *possible, proposed, expected, doubtful*. (Partee, 2007, p. 9)

Similarly, Kamp (1975) notes that adjectives like 'alleged', 'fake', 'skillful', 'good' resist his attempt, *contra* (Montague, 1970), to reduce the meaning of most adjectives from property-to-property mappings to properties. It is well-known that there are English adjectives which cannot occur in predicative position and must occur in attributive position, e.g. 'mere'. And the converse holds for some English adjectives, like 'asleep'. Paoli (1999, p. 69) adduces the example "The former president is asleep", which does

---

[1] This research has been supported by the Internal Grant Agency of VŠB-TU Ostrava, Project No. SP2013/207.

not convert to *"The asleep president is former". See also (Beesley, 1982, pp. 205–209) and (Bolinger, 1967). I suggest that adjectives that can occur only in attributive position must denote property modifiers and cannot denote properties. Thus in 'alleged assassin' the adjective 'alleged' denotes a modal modifier while the noun 'assassin' denotes a property and the noun phrase 'alleged assassin' denotes a modified property. The sentence "$a$ is an alleged assassin" is not a telescoped variant of the conjunction *"$a$ is alleged and $a$ is an assassin". The first conjunct is ill-formed, and the second conjunct pre-empts the open question whether $a$, who is an alleged assassin, is indeed an assassin.

Partee is right, of course, that an alleged murderer may or may not be a murderer. But she is not right to infer from this observation that modal modifiers/adjectives have no entailments. They do have—non-trivial—entailments.[2] This is good news, anyway, for Partee's theory of adjectives. If modal modifiers were to lack a meaning postulate for want of entailments, then how could her *non-vacuity principle* accommodate modal modifiers?[3] That principle prescribes that the semanticist interpret any predicate in such a manner that both its positive and negative extension come out non-empty. The non-vacuity principle is in danger of being rendered inapplicable to a perfectly normal and natural adjective like 'alleged' or noun phrase like 'alleged assassin' for want of a meaning postulate to control modal adjectives / modifiers. One might object that 'alleged' cannot have an extension, either positive or negative, simply because modifiers are not the sort of thing that can have a satisfaction class (they are not true or false of anything). So maybe the non-vacuity principle was never intended for predicates or adjectives that cannot occur in predicative position. But the principle should apply to 'alleged assassin'. How, though?

The challenge that the existence of modal modifiers presents is to carve out a niche for them in the overall edifice of modifiers. The present paper crystallizes the logical behaviour that is unique to modal modifiers. Their behaviour will be captured in a definition, and this definition will underlie an introduction and an elimination rule controlling modal property and propositional modifiers, respectively. That is, the meaning of an adjective denoting a property modifier and the meaning of an adverb denoting a propositional modifier is made explicit by a definition and not by an introduction rule. My

---

[2] Partee stresses that she did not intend her 2007 quote to suggest that modal adjectives lack meaning altogether (personal communication, *LOGICA* 2012, Hejnice, 18-22 June 2012).

[3] I owe this observation to Michael Johnson. The principle is stated in (Kamp & Partee, 1995, p. 161).

semantics will be worked out within Transparent Intensional Logic. TIL is a three-tiered theory, with one tier for extensional entities, one for intensional entities (as understood by possible-world semantics), and one for hyperintensional entities. The entities of the first two tiers are organized into a simple type theory. The entities of the last tier are organized into a ramified type hierarchy. My analysis of modifiers requires only intensional entities, like properties and propositions, and mappings defined over intensional entities. Modification does not require us to go hyperintensional. The fragment of TIL within which I have developed the theory of modification is continuous with the general model-theoretic one Montague, Kamp, Partee and others have conceived. The TIL theory of modal modification fills a gap in that theoretical edifice.

The sample sentences I will be investigating here are "$a$ is an alleged assassin" and "Allegedly, $a$ is an assassin". The approach I am fielding here is in line with (Jespersen & Carrara, 2011, to appear), which analyze "$a$ is a malfunctioning $F$", 'malfunctioning' denoting either a subsective or a privative modifier; (Carrara & Jespersen, in submission), which analyzes iterated modification, as in "$a$ is fake rhinestone jewelry"; and (Primiero & Jespersen, 2010), which analyzes "$a$ is a fake banknote". The present paper draws and expands upon material found in (Primiero & Jespersen, 2013), which analyzes "$a$ is an alleged assassin" and "Allegedly, $a$ is an assassin". Particularly the definitions of property and propositional modal modifiers have been stated slightly more carefully here, and so have the two elimination rules.

## 2 The Edifice of Property Modifiers

Property modifiers typically divide into four kinds. This is how I prefer to characterize them, $[MF]_{wt}$ being the set of individuals (an extensionalized property) with the modified property $[MF]$ at the arbitrary world-time couple $\langle w, t \rangle$:

- *Subsective.* $[M_s F]_{wt} a \therefore F_{wt} a$ (e.g. a *large* horse is a horse).

- *Privative.* $[M_p F]_{wt} a \therefore [nonF]_{wt} a$ (e.g. a *fake* banknote is a non-banknote).

- *Modal.* $[M_m F]_{wt} a \therefore$ see below (e.g. an *alleged* assassin is maybe an assassin).

- *Intersective.* $[M_i F]_{wt} a \therefore M^*_{wt} a \wedge F_{wt} a$ (e.g. a *round* peg is round and a peg).

Remark 1. A limiting case of subsective modification is constituted by semantically and logically *trivial* or *redundant* modifiers, like *real* or *genuine*.[4] Trivial modifiers return their arguments unmodified, so to speak. A trivial modifier is the identity function defined over properties, taking properties to themselves: hence $\forall \lambda x[[\mathit{Genuine} F]_{wt} x \equiv [F_{wt} x]]$. Since subset and superset coincide, trivial modifiers are neither upward nor downward monotonic. However, the non-trivial from among the subsective modifiers are upward monotonic. Terms for identity functions lack logical and semantic import, but they may have rhetoric or perlocutionary import, as when bringing out the contrast between what is fake and what is genuine. Or think of the Paschal greeting: the first speaker says, "Christ is risen!" and the other responds, "Truly He is risen!", 'truly' serving to emphasize the second speaker's enthusiastic agreement with what the first speaker just said. See (Primiero & Jespersen, 2013, n. 5), (Jespersen & Carrara, to appear, n. 13) on *genuine*, and (Duží, Jespersen, & Materna, 2010, pp.189–190) on *actually* as the identity function over logical space. Note that introducing a non-empty category of trivial modification clashes with the non-vacuity principle. Her adherence to the principle saddles (Partee, 2001) with an ontological dichotomy between two kinds of fur, guns, etc.: some fur is fake, some fur is real, some guns are fake, some guns are real, etc. I much prefer housing the fake/genuine dichotomy exclusively on the linguistic level.

Remark 2. The *general* rule I recommend for privative modification is $[M_p F]_{wt} a \therefore [\mathit{non} F]_{wt} a$, where *non* is a property-to-property function. In the case of *single* privation, the general rule is equivalent, but not identical, with the specific one, which is stronger: $[M_{p'} F]_{wt} a \therefore \neg F_{wt} a$. The specific rule replaces $M_{p,}$, which operates on properties, by $\neg$, which operates on truth-values (or else propositions, as in constructivist conceptions of negation). The basic reason for replacing $\neg$ by *non* is that if double privation replaces two instances of a privative modifier by two instances of $\neg$ then we obtain $\neg\neg [F_{wt} a]$, classically equivalent to $[F_{wt} a]$. *E voilá* fake rhinestone jewelry emerges as jewelry! Our resulting logic of double privation is a logic of contraries rather than contradictories. A pair of privative modifiers is tantamount to one modal modifier. Hence fake rhinestone jewelry is

---

[4]'Real' has a different meaning in a sentence such as, "That's a real disaster!". Here 'real' is a non-trivial subsective modifier selecting those disasters that qualify as serious disasters (against the background of a scale of gravity for disasters).

maybe jewelry. (Carrara & Jespersen, in submission, §3) has the details.

Remark 3. If all modifiers were subsective then in $[MF]$ property $F$ would be the *genus* and $M$ one of the *differentia* qualifying this genus, as in [*Herbivorous Mammal*]. But privative modifiers upset this neat setup. Fake banknotes are not a subspecies of banknotes, nor are alleged assassins, always, a subspecies of assassins, though occasionally they are. Yet there is a derivative sense in which privative modifiers are subsective. Primiero and Jespersen (2010, §4.2) argues that a fake banknote, say, is not just any old non-banknote, but a member of a well-defined subset of the *complement* of any set of banknotes. That is, grab property $F$, extensionalize it to obtain the set $F_{wt}$, apply the function $\setminus$ to obtain its complement, and finally obtain the subset $[Fake_p F]_{wt}$ by applying the condition $[Fake_p F]$ to the elements of $\setminus F_{wt}$. This suggestion goes some way at least toward accommodating Partee's project of reducing all modifiers, bar some modal ones, to subsective modifiers: (Partee, 2001, p. 7).

Remark 4. Intersective modification needs a rule of right subsectivity to infer the second conjunct (i.e. the rule of subsective modification) and a rule of left subsectivity to infer the first conjunct. (Duží et al., 2010, §4.4), (Jespersen & Carrara, to appear, §2.2), Jespersen (in submission) launch pseudo-detachment as a rule of left subsectivity to replace the modifier $M$ in the premise by the property $M^*$ in the conclusion. The idea is that if $a$ has the modified property $[MF]$, $M$ any kind of modifier, then there is a property $p$ such that $a$ has property $[Mp]$. Abbreviate '$[Mp]$' as '$M^*$'. Then the proper statement of the rule of intersection turns out to be

$$\frac{[M_i F]_{wt} a \quad [M_i F]_{wt} a}{M^*_{wt} a \quad F_{wt} a}$$
$$\overline{M^*_{wt} a \wedge F_{wt} a}$$

Partee (2001, p. 3) puts forward a meaning postulate to control intersective adjectives: For each intersective meaning $ADJ'$ it holds that

$$\exists P_{\langle e,t \rangle} \Box \forall Q_{\langle s, \langle e,t \rangle \rangle} \forall x_e [ADJ'(Q)(x) \leftrightarrow P(x) \wedge {}^\vee Q(x)]$$

The meaning postulate gets the truth-condition right, for sure, but fails to account for the *logic* validating the transition from $ADJ'$ to $P$, in the absence of an explicitly stated rule of left subsectivity.

## 3  The Unique Features of Modal Modifiers

Modal modification is readily characterized *negatively*. $[M_m F]_{wt} a$ *fails* to validate either of $F_{wt} a$, $[nonF]_{wt} a$ as conclusion and so $M_m$ is *non*-committal. Hence modal modifiers oscillate between subsection and privation.[5] $M_m$ is *non*-intersective for failure to obey the rule of intersective modification. Therefore, $M_m$ is *non*-extensional for failure to obey this rule (cf. Kamp, 1975, pp. 125–126):

$$\frac{F_{wt}x \leftrightarrow G_{wt}x}{[M_e F]_{wt}x \leftrightarrow [M_e G]_{wt}x}$$

Extensional modifiers are *downward monotonic*: if $F_{wt}, G_{wt}$ are the same set, then $[M_e F]_{wt}, [M_e G]_{wt}$ are the same subset within $F_{wt}$. So it makes no logical difference whether an element of $F_{wt}$ has $[M_e F]_{wt}$ or $[M_e G]_{wt}$ predicated of it.

What would a non-trivial, *positive* characterization of modal modifiers look like? It is obvious that it would be a non-starter to simply slam together the respective conclusions of subsective and privative modification by means of disjunction: $[M_m F]_{wt} a \therefore F_{wt} a \lor [nonF]_{wt} a$. That conclusion is just a classical tautology, hence satisfied by all the other modifiers as well. Necessarily, subsective modifiers satisfy the left-hand disjunct and, necessarily, privative modifiers satisfy the right-hand disjunct. So the conclusion $F_{wt} a \lor [nonF]_{wt} a$ is a positive one, but also trivial. Note, however, that the rule of pseudo-detachment validates this inference:

$$\frac{[alleged\ assassin]_{wt} a}{\exists p[[alleged p]_{wt} a]}$$

Although pseudo-detachment applies to all four kinds of modifiers, it is non-trivial to be alleged to have this, that or the other property. Some people are not at the receiving end of an allegation. So this is one interesting inference/entailment that modal modifiers do have.

For a more general account, here is a suggestion inspired by the analysis of *knowing whether*, which is a non-factive attitude, found in (Duží

---

[5] Partee (2007, pp. 151–152) lists 'alleged' among the 'plain non-subsective' adjectives, but that cannot be right, for in each individual case where an alleged $F$ is indeed an $F$ 'alleged' is subsective. Abdullah and Frost (2005, p. 330) lists 'alleged' as both 'non-subsective' and 'non-privative', which also cannot be right, but at least their double deprivation gesticulates toward the oscillation between subsection and privation.

et al., 2010, §5.1.4). The suggestion is this. When the proposition $\lambda w \lambda t[[M_m F]_{wt} a]$ is true at $\langle w, t \rangle$ then one of two *alethic* and *epistemic possibilities* is realized at $\langle w, t \rangle$: $a$ being an $F$; $a$ being a non-$F$. If $a$ is an alleged assassin at some $\langle w, t \rangle$ then it is alethically and epistemically possible that $\langle w, t \rangle$ be identical to some $\langle w', t' \rangle$ at which $a$ is an assassin, and it is likewise possible that $\langle w, t \rangle$ be identical to some $\langle w'', t'' \rangle$ at which $a$ is a non-assassin. The open question, alethically and epistemically, is whether $\langle w, t \rangle$ is like $\langle w', t' \rangle$ or like $\langle w'', t'' \rangle$. $M_m$ behaves with respect to one and the same property $F$ as subsective at $\langle w', t' \rangle$ and as privative at $\langle w'', t'' \rangle$.[6]

None of the three other modifiers has the context-sensitive feature that its status (here: subsective or privative) depends on the given $\langle w, t \rangle$ of evaluation. To contrast modal with subsective and privative modifiers, it is necessary that if $\lambda w \lambda t[[M_s F]_{wt} a]$ is true then $\lambda w \lambda t[F_{wt} a]$ follows, and it is necessary that if $\lambda w \lambda t[[M_p F]_{wt} a]$ is true then $\lambda w \lambda t[[nonF]_{wt} a]$ follows. What is unique about modal modifiers is that it will not do to investigate the logical relationship between modifiers and properties to establish whether an individual instantiating some modified property $F$ has property $F$. Rather we must investigate, empirically, each individual instantiation of $[M_m F]$ to establish whether an individual having property $[M_m F]$ also has property $F$. There are exceptions on the periphery, of course. If $F$ is the property of being a clean-shaven barber such that he shaves all and only those who don't shave themselves then, unless we're paraconsistent logicians, there's no need to jump out of the armchair and into the field to establish that $F$ isn't true of anyone of whom $[M_m F]$ is true.

## 4 A Formal Semantics for Modal Modifiers

Below I work out in formal detail the proposal to analyze modal modification in terms of possibility.

We first define the simple type theory.

**Definition 1** (types of order 1 over $B$)  *Let B be an ontological base, i.e. a collection of pair-wise disjoint, non-empty sets. Then:*

*(i) Every member of B is an elementary type of order 1 over B.*

---

[6]Daley (2010) attempts to apply TIL to, e.g., 'alleged felon'. To avoid inferring from $x$ being an alleged felon that $x$ is a felon, Daley (2010, pp. 366–367) assigns one type to *felon* when occurring alone and another type when occurring together with *alleged*. Type-shifting is not an option in TIL, however. Besides Daley's type assignments are incoherent. See (Jespersen, in preparation) for further details and corrections.

(ii) Let $\alpha, \beta_1, \ldots, \beta_n$ ($n > 0$) *be types of order* 1 *over B. Then the collection* $(\alpha\beta_1 \ldots \beta_n)$ *of all n-ary partial mappings from* $\beta_1 \times \ldots \times \beta_n$ *into* $\alpha$ *is a functional type of order* 1 *over B*.

(iii) *Nothing is a type of order* 1 *over B unless it so follows from (i), (ii)*.

Remark 5. The ground types currently encompass these four:

$o$ the set of truth-values $\{\top, \bot\}$,

$\iota$ the universe of discourse (individuals),

$\tau$ the set of reals, doubling as times,

$\omega$ the logical space of logically possible worlds.

Remark 6. Intensional entities belong to the functional types with domain in $\omega$. A *proposition* is a function of type $((o\tau)\omega)$, abbreviated '$o_{\tau\omega}$', from worlds to a partial function from times to truth-values. A *property* is a function of type $(((o\iota)\tau)\omega)$, abbreviated '$(o\iota)_{\tau\omega}$', a function from worlds to a function from times to sets of individuals. See (Duží et al., 2010, §1.4.2.1) for a classification of properties.

Remark 7. *Modifiers* are not intensional, but extensional, entities. They are functions from one intension to another. A *property modifier* is of type $((o\iota)_{\tau\omega}(o\iota)_{\tau\omega})$, forming a property from a property. A *propositional modifier* is of type $(o_{\tau\omega}o_{\tau\omega})$, forming a proposition from a proposition.

Remark 8. I am making a type-theoretic and notational shortcut. The formula '[*modifier property*]' dressed up in full TIL notation and with appropriate type-theoretic assignments should be written as '$[^0 modifier\ ^0 property]$'. $^0 property$, $^0 modifier$ are higher-order objects of the lowest order, which are modes of presentation, or *constructions*, of first-order objects such as modifiers and properties. The shortcut saves me from a detour around constructions, which are only being used here, anyway, to present the sort of objects I want to operate on and are not objects of study in their own right. See (Duží et al., 2010, §1.3.2) on constructions and (*ibid.*, §2.6.1), (Duží & Jespersen, 2012; Jespersen & Duží, 2013) on the use/mention distinction as applied to constructions.

The definition of modal modifiers below defines the *set* of modifiers $g$ that are modal with respect to $F$, such that if $a$ is a $[g_m F]$ at $\langle w, t \rangle$ then at $\langle w, t \rangle$, possibly, $a$ is an $F$ and, possibly, $a$ is a non-$F$. The reason for making the status of the value of $g$ depend on its argument property $F$ is that I don't want to assume that it be fixed for each modifier whether it is subsective or privative or modal or intersective. For instance, *fake* will typically be considered privative, but is *fake information* non-information or information that is untrue? I want to leave conceptual space for the second option as well. Similarly, where an alleged assassin may, or may not, turn out to be an assassin, an *alleged proposition* is a proposition that has been alleged as true.

An auxiliary definition of *requisite* is required: see (Duží et al., 2010, §4.1). A requisite is, formally, a relation-in-extension between two intensions. For instance, if $F, G$ are properties and $G$ a requisite of $F$ then, necessarily, anything that has $F$ also has $G$. We need the requisite relation in order to express that (i) necessarily, if $x$ has some particular modified property then $x$ also has some particular plain property, and (ii) necessarily, if proposition $P$ is true then proposition $Q$ is also true. The requisite relation obtaining between propositions coincides with the relation of entailment when restricted to a one-membered premise set.

**Definition 2** (requisite of property)  *Let $\forall/(o(o\alpha))$ be a function from a set of $\alpha$-objects (e.g. individuals, worlds, times) to a truth-value; let $X, Y$ range over $(o\iota)_{\tau\omega}$, $x$ over $\iota$, $w$ over $\omega$, and $t$ over $\tau$. Then $[ReqYX] =_{df} \forall \lambda w [\forall \lambda t [\forall \lambda x [X_{wt}x] \supset [Y_{wt}x]]]$.*

Gloss *definiendum* as "$Y$ is a requisite of $X$" and *definiens* as "Necessarily, any $x$ instantiating $X$ at any $\langle w,t \rangle$ also instantiates $Y$ at $\langle w,t \rangle$". Hence, if at $\langle w,t \rangle$ $a$ has the modified property $[M_s F]$ then it is necessary that $a$ also have the property $F$ at $\langle w,t \rangle$: necessarily, whatever is an $[M_s F]$ is an $F$.

Remark 9. In case $[X_{wt}x]$ in $[X_{wt}x] \supset [Y_{wt}x]$ is *improper*, for failure to produce a value, the right-hand side of $=$ in Def. 2 won't be true while the left-hand side will. (Jespersen & Carrara, to appear, §2.4) shows how a truth predicate restores equality. Def. 2 disregards improper constructions and partial functions in the interest of simplicity.

**Definition 3** (subsective, privative, modal, intersective property modifiers)  *Let $M/((o\iota)_{\tau\omega}(o\iota)_{\tau\omega})$; let $g^*$ range over $(o\iota)_{\tau\omega}$; let $g_s, g_p, g_m, g_i$ range over $((o\iota)_{\tau\omega}(o\iota)_{\tau\omega})$; let $F/(o\iota)_{\tau\omega}$; let $x$ range over $\iota$; let $\langle w',t' \rangle \neq \langle w'',t'' \rangle$; let $\in /(o((o\iota)_{\tau\omega}(o\iota)_{\tau\omega})(o((o\iota)_{\tau\omega}(o\iota)_{\tau\omega})))$, i.e. it is true or false that a particular modifier is an element of a particular set of modifiers. Then:*

*$M$ is subsective w.r.t. $F$ iff $M \in \lambda g_s [ReqF[g_s F]]$,*

$M$ is privative w.r.t. $F$ iff $M \in \lambda g_p[Req[nonF][g_pF]]$,

$M$ is modal w.r.t. $F$ iff
$M \in \lambda g_m[Req[\lambda w \lambda t[\lambda x[[\exists \lambda w'[\exists \lambda t'[[[M_mF]_{wt}x] \to [F_{w't'}x]]]] \wedge [\exists \lambda w''[\exists \lambda t''[[[M_mF]_{wt}x] \to [[nonF]_{w''t''}x]]]]]][g_mF]]]$,

$M$ is intersective w.r.t. $F$ iff $M \in \lambda g_i[Req[\lambda w \lambda t \lambda x[[g^*_{i_{wt}}x] \wedge [F_{wt}x]]][g_iF]]$.

The property modifier *alleged* can be defined thus:

**Definition 4** (alleged)  Let $='/(o(o\iota)_{\tau\omega}(o\iota)_{\tau\omega})$; let $\imath/((o\iota)_{\tau\omega}o(o\iota)_{\tau\omega})$: this singularizer takes a singleton of properties to its element and is otherwise undefined; let $f, f'$ range over $(o\iota)_{\tau\omega}$; let $allege/(o\iota o_{\tau\omega})_{\tau\omega}$ be a relation-in-intension between individuals and propositions they allege to be true. Then

$alleged =_{df} \lambda f[\imath \lambda f'[f' ='$
$=' \lambda w \lambda t[\lambda y[\exists \lambda x[alleges_{wt}x\lambda w' \lambda t'[f_{w't'}y]]]]]]$.

Remark 10. *Alleged* is the property modifier that takes a property $f$ to the property $f'$ of being someone, $y$, whom somebody, $x$, alleges to be an $f$. Thus $f'$ is the property of being an alleged $f$.

The *introduction rule* for the modifier *alleged* is straightforward now, on the simplifying assumption that allegations are relations to possible-world propositions rather than hyperpropositions:

$$\frac{\exists x[alleges_{wt}x\lambda w \lambda t[f_{wt}a]]}{[alleged f]_{wt}a}$$

Remark 11. A *speech act* consisting in somebody making an allegation that reveals a certain *attitude* to a particular possible state-of-affairs is conceptually prior to $a$ acquiring the property of being an alleged $f$. (De-Lazero (2011) argues, not implausibly, that modal adjectives/modifiers basically modify events or states, not properties of individuals.) I'm not sure, at this point, how to extend the introduction rule for *alleged* and other *attitude*-dependent modal modifiers to the entirety of all modal modifiers. Hence I'm not offering, at this point, a general introduction rule for modal property modifiers.

Remark 12. I am disregarding, in the interest of simplicity, examples of allegations that involve *partiality*, as expressed by "$b$ alleges that the King of

Denmark is a Frenchman". (Cf. Remark 9.) The proposition that the King of Denmark is a Frenchman currently returns a truth-value gap, because the individual office of King of Denmark/$\iota_{\tau\omega}$ is currently vacant. Still it is either true or false that $b$ has made/is making such an allegation. See (Duží et al., 2010, pp. 276–278) on the logic and philosophy of partiality.

From Def. 3 we obtain the following approximation to an *elimination rule* for $M_m$, $f$ ranging over properties:

$$\frac{[M_m f]_{wt} a}{\exists \lambda w'[\exists \lambda t'[[[M_m f]_{wt} a] \to [f_{w't'} a]]] \land \exists \lambda w''[\exists t''[[[M_m f]_{wt} a] \to [non f]_{w''t''} a]]}$$

Gloss: "From $a$ being an $[M_m f]$ at $\langle w, t \rangle$, infer that there is a $\langle w', t' \rangle$ such that if $a$ is an $[M_m f]$ at $\langle w, t \rangle$ then $a$ is an $f$ at $\langle w', t' \rangle$ and that there is a different $\langle w'', t'' \rangle$ such that if $a$ is an $[M_m f]$ at $\langle w, t \rangle$ then $a$ is a *non-f* at $\langle w'', t'' \rangle$."

Absolute elimination of $M_m$ in the conclusion is impossible due to the oscillation between subsection and privation, so the proposed rule makes do with conditional elimination.

Let *allegedly* be a propositional modifier. Then the analysis of the sentence "Allegedly, $a$ is an assassin", or "$a$ is allegedly an assassin", is $[Allegedly \lambda w \lambda t[Assassin_{wt} a]]$: see (Primiero & Jespersen, 2013, §2.3).

It suffices to provide the definition of the modal from among the propositional modifiers here. The definitions of the remaining propositional modifiers are easily reconstructed.

**Definition 5** (modal propositional modifier)  *Let $M'/(o_{\tau\omega}o_{\tau\omega})$; let $g'$ range over $(o_{\tau\omega}o_{\tau\omega})$; let $P/o_{\tau\omega}$. Then*

$M'$ *is modal w.r.t.* $P$ *iff* $M' \in \lambda g'_m[Req[\lambda w \lambda t[\exists \lambda w'[\exists \lambda t'[g'_m P]_{wt} \to P_{w't'}]] \land [\exists \lambda w''[\exists \lambda t''[g'_m P]_{wt} \to \neg P_{w''t''}]][g_{m'} P]]$.

The definition of *allegedly* mirrors the definition of *alleged*:

**Definition 6** (allegedly)  *Let $='' /(oo_{\tau\omega}o_{\tau\omega})$; let $p, p'$ range over $o_{\tau\omega}$; let $\jmath'/(o_{\tau\omega}(o(o_{\tau\omega})))$. Then*

*allegedly* $=_{df} \lambda p[\jmath' \lambda p'[p' ='' \lambda w \lambda t[\exists \lambda x[alleges_{wt} xp]]]]$.

The *introduction rule* for *allegedly* is straightforward:

$$\frac{\exists \lambda x[alleges_{wt} xP]}{[allegedly P]_{wt}}$$

The corresponding conditional *elimination rule* is

$$\frac{[allegedlyP]_{wt}}{\exists \lambda w'[\exists \lambda t'[[[allegedlyP]_{wt}] \to [P_{w't'}]]] \wedge} \\ \exists \lambda w''[\exists \lambda t''[[[allegedlyP]_{wt}] \to [\neg P_{w''t''}]]]$$

Provided '$a$' denotes an individual and not an individual office, "Allegedly, $a$ is an assassin" and "$a$ is an alleged assassin" come out equivalent sentences by denoting the same truth-condition (possible-world proposition), but they are not synonymous. In the vernacular of TIL we say that the (value-forming construction) Composition [$^0$*Allegedly* $\lambda w \lambda t$ [$^0$*Assassin*$_{wt}$ $^0a$]] and the (function-forming construction) Closure $\lambda w \lambda t$[[$^0$*Alleged* $^0$*Assassin*]$_{wt}$ $^0a$] are equivalent, in accordance with Def. 1.5 in (Duží et al., 2010, p. 48), but not procedurally isomorphic, as defined by Def. 2.3 (*ibid.*, p. 154). See (Jespersen & Duží, 2013) for the latest definition of procedural isomorphism.

## ACKNOWLEDGMENTS

Versions or portions of this paper were read at *TbiLLC 2011*, 26-30 September 2011, Kutaisi, Georgia; Department of Philosophy, Hong Kong University, 8 May 2012; Department of Chinese and Bilingual Studies, Hong Kong Polytechnic University, 7 May 2012; *LOGICA 2012*, 18-22 June 2012. I wish to thank the audiences, Michael Johnson and Dan Marshall in particular, for vigorous discussion. I also wish to thank Marie Duží for very helpful suggestions, and Giuseppe Primiero for great cooperation on our two published papers on modifiers.

## References

Abdullah, N., & Frost, R. (2005). Adjectives: A uniform semantic approach. *Lecture Notes in Artificial Intelligence, 3501*, 330–341.

Beesley, K. R. (1982). Evaluative adjectives as one-place predicates in Montague Grammar. *Journal of Semantics, 1*, 195–249.

Bolinger, D. (1967). Adjectives in English: attribution and predication. *Lingua, 18*, 1–34.

Carrara, M., & Jespersen, B. (in submission). Double privation and multiply modified artefact properties.

Daley, K. (2010). The structure of lexical concepts. *Philosophical Studies*, *150*, 349–372.

DeLazero, O. E. (2011). On the semantics of modal adjectives. *University of Pennsylvania Working Papers in Linguistics*, *17*. (Available at: http://repository.upenn.edu/pwpl/vol17/iss1/11)

Duží, M., Jespersen, B., & Materna, P. (2010). *Procedural semantics for hyperintensional logic: Foundations and applications of Transparent Intensional Logic, LEUS* (Vol. 17). Berlin: Springer-Verlag.

Duží, M., & Jespersen, B. (2012). Transparent quantification into hyperpropositional contexts *de re*. *Logique et Analyse*, *220*, 513–554.

Jespersen, B. (in preparation). Structured lexical concepts, property modifiers, and Transparent Intensional Logic: reply to Daley.

Jespersen, B. (in submission). Left subsectivity.

Jespersen, B., & Carrara, M. (2011). Two conceptions of technical malfunction. *Theoria*, *77*, 117–138.

Jespersen, B., & Carrara, M. (to appear). A new logic of technical malfunction. *Studia Logica*.

Jespersen, B., & Duží, M. (2013). Procedural isomorphism, analytic information and $\beta$-conversion by value. *Logic Journal of the IGPL*, *21*, 291–308.

Kamp, H. (1975). Two theories of adjectives. In E. Keenan (Ed.), *Formal semantics of natural language* (pp. 123–155). Cambridge: Cambridge University Press.

Kamp, H., & Partee, B. (1995). Prototype theory and compositionality. *Cognition*, *57*, 129–191.

Montague, R. (1970). English as a formal language. In Visentini et al. (Ed.), *Linguaggi nella societá e nella tecnica* (pp. 189–224). Milan.

Paoli, F. (1999). Comparative logic as an approach to comparison in natural language. *Journal of Semantics*, *16*, 67–96.

Partee, B. (2001). Privative adjectives: subsective plus coercion. In R. Bäuerle, U. Reyle, & T. Zimmermann (Eds.), *Presupposition and discourse*. Amsterdam: Elsevier.

Partee, B. (2007). Compositionality and coercion in semantics: The dynamics of adjective meaning. In G. Bouma et al. (Ed.), *Cognitive foundations of interpretation* (pp. 145–161). Amsterdam: Royal Netherlands Academy of Arts and Sciences.

Primiero, G., & Jespersen, B. (2010). Two kinds of procedural semantics for privative modification. *Lecture Notes in Artificial Intelligence*, *6284*,

252-271.

Primiero, G., & Jespersen, B. (2013). Alleged assassins: realist and constructivist semantics for modal modifiers. *Lecture Notes in Computer Science*, *7758*, 94–114.

Bjørn Jespersen
Department of Computer Science
VŠB-Technical University of Ostrava
Ostrava, Czech Republic
e-mail: `bjornjespersen@gmail.com`

# On Modal Facts in Possible Worlds

## Neil Kennedy[1]

**Abstract:** In this paper, I will focus on certain limitations of the possible worlds semantic framework. These limitations stem from the impossibility (or near impossibility) of individuating modal facts. The implications of this are twofold. First, it will prove difficult to provide a "constructive" or "iterative" account of worlds, that is, an account where worlds are built from the bottom-up out of more basic constituents. Second, and related to the first point, the absence of modal parts will make the semantics of certain types of higher-order modal statements elusive. The positive contribution of this paper will be a framework in which modal facts are part of the worlds themselves.

**Keywords:** possible worlds, modality

## 1 Possible Worlds and Modal Facts

Possible worlds have become the standard machinery for the analysis of modality. Using them as a basic semantic ingredient, we can provide modal expressions of different creeds with compositional meanings. In the simplest (and most common) cases, modals are analyzed as restricted quantifiers over a set of possible worlds, where the restriction typically takes the form of a binary accessibility relation. In more complicated cases, a modal expression may require more than an accessibility relation,[2] but the general idea will remain the same: the truth conditions of a modal statement are specified using a set of worlds $W$ and some additional structure $\mathbf{S}$ on top of it.

Though this framework has many theoretical virtues, in what follows, I will argue that it also presents some serious limitations. These limitations stem from the impossibility (or near impossibility) of individuating modal facts, as the latter only exist as one indivisible whole in the structural component $\mathbf{S}$ of the model. The implications of this are twofold. First, it will prove difficult to provide a "constructive" or "iterative" account of worlds,

---

[1] This work is supported by grant # 149410 of the FQRSC. I would like to thank audiences at *Trends in Logic 2012* and *Logica 2012*, and I'm grateful for comments from Agustin Rayo.

[2] For example, the Stalnaker-Lewis semantics for counterfactual conditionals presuppose a world bound similarity relation on worlds.

that is, an account where worlds are built from the bottom-up out of more basic constituents. Second, and related to the first point, the absence of modal parts will make the semantics of certain types of higher-order modal statements elusive. The positive contribution of this paper will be a framework in which modal facts are part of the worlds themselves.

## 2 The Locus of Modal Facts

The "truthmaking" or "truth-value-determining" powers of a Kripke model $\langle W, \mathbf{S} \rangle$ have two distinct aspects to them. First of all, possible worlds, when taken in isolation of the structure $\mathbf{S}$, can be seen as maximal arrangements of non-modal facts. This is basically because the truth value of any (non-modal) atomic sentence at any given world is entirely determined by that world. What distinguishes worlds from other truthmaking (or truth-value-determining) entities is the exhaustive nature of the truthmaking: situations and states of affairs also determine the truth values of atomic sentences, but they may fail to determine them all.

However, it is clear that possible worlds alone, if taken in isolation from the structure $\mathbf{S}$, will not suffice to determine the truth values of modal statements. It is only in conjunction with $\mathbf{S}$ that a possible world can determine the truth value of all statements (modal and non-modal alike). If modal facts are the kind of things that help make modal statements true (or false), then it is clear from the previous observation that modal facts aren't located in any possible world, but are somewhat spread out and distributed over $W$ and $\mathbf{S}$.

An analogous observation can be made about probability spaces, as they share many things in common with possible worlds. The points of a probability space can be seen as maximal configurations of "non-probabilistic" facts: they settle everything there is to settle that isn't probabilistic in nature, e.g. the values of the dice, the colours of the balls chosen from the urns, etc. The probabilistic facts, however, are located at the structural level, with the probability distribution as the main structuring element of the space.

The distributed nature of modal (and probabilistic) facts means, quite simply, that modal facts aren't parts or constituents of worlds. If modal facts live over and above individual worlds, then it will be hard to give a bottom-up account of possible worlds that "generates" the model $\langle W, \mathbf{S} \rangle$. This would constitute a major setback for any conception of worlds that subscribes to the principle of recombination (cf. Lewis, 1986, pp. 87–88; Armstrong, 1989, pp. 20–23), since modal facts will not be so readily available to combine. But more importantly, I think, this feature makes the semantics ill-suited for certain types of higher-order modality. Some higher-order modal combinations, I would argue, are more naturally understood as quantifiers over modal facts. In the Kripkean setting, it will be impossible to directly represent their meaning as such.

## 3  World Building In Kripke Semantics

We illustrate these points with a series of examples. Suppose we have a universe $U$ consisting of a single tree that can bear one of only three properties: that of being a maple tree, of being an oak tree, or an elm tree. Hence, $U$ has three possible states, one for each possible property the individual can have. Now add Alice to $U$, and suppose that the only property she has is that of being a tree-expert, where a tree-expert is someone who can distinguish maples from oaks, maples from elms and oaks from elms. Tree-expertise of this kind constitutes a modal (epistemic) fact about Alice. How do we go about adding this modal fact to $U$?

A natural suggestion would be to fuse Alice and her tree-expertise to each of the three previous tree states. The resulting universe, say $U'$, would still consist of three possible states, but the states will now also pertain to Alice's tree-expertness. In particular, we will expect each of these new states to make the following true:

(1)  Alice is a tree-expert

Moreover, we will also expect that the truth of (1) entails the truth of

(2)  If it's a maple tree, then Alice will know it (if she can examine it properly, etc.)
(3)  If it's an oak tree, then Alice will know it
(4)  If it's an elm tree, then Alice will know it

Unfortunately, there is nothing in the new states of $U'$ that forces statements (2)-(4) to be true, the problem being that the property of being a tree expert isn't hooked up to Alice's knowledge in any way.

In the context of Kripke semantics, a most expedient solution to this difficulty would be to treat (1) not as a report of some epistemic property holding of Alice but as shorthand for (2)-(4), so that (1) would follow by definition and from the nature of Alice's total knowledge. Hence, the transition from the old universe $U$ to the new universe $U'$ would not be accomplished by fusing Alice and her tree-expertness to possible states of $U$, but by adding structure on top of the set of possible states of $U$, namely, by adding an accessibility relation $R$ on these states that translates her tree discrimination powers. There is no doubt that this makes the entailments true, but the nature of this added relation or structure still remains to be explained.

The specific nature of $R$ is made clearer when other facts are added to $U'$. Suppose the tree in the universe has leaves and that these leaves can be either green, yellow or brown; assume, to simplify, that they are all one and the same colour. The set of possible states of the new universe, say $U''$, are obtained from the possible states of $U'$ by "fusing" each one of these states with each possible colour state of the leaves. Even if we assume that Alice has the same tree-expertise in $U''$, $R$ will no longer represent it correctly. For one, $R$ isn't defined on the set of possibilities of $U''$. But suppose we "update" $R$ in some way or other to make it a relation $R''$ on the set of possibilities of $U''$. Whichever way we do this, the problem is that $R''$ will no longer just pertain to tree-expertise but it will also pertain to colour-expertise. In other words, this relation will automatically determine the truth value of a statement such as

(5) Alice is red-green colourblind

if we understand the latter as shorthand for

(6) If the leaves are red or green, then Alice won't know which one it is
(7) If the leaves are brown, then Alice will know it (if she can examine them, etc.)

What the example illustrates is that these modal (epistemic) facts concerning Alice can't be added incrementally, they are given as one indivisible whole in the accessibility relation.

Another difficulty lies in modal (epistemic) facts that concern other modal facts. Suppose we now add Bob to the universe, and assume that the following is true of his knowledge of Alice:

(8) For all Bob knows, Alice might be red-green colourblind or colour-seeing

(We can assume whatever we want about his knowledge of trees and colours, it won't matter for the following.) (8) is true whenever there exist two possible states $w$ and $v$ such that Alice has the property of being red-green colourblind in $w$ and the property of being colour-seeing in $v$. As we have emphasized, the truthmaking difference between $w$ and $v$ resides not in there being something in $w$ not in $v$ and vice versa (namely a red-green colourblind Alice in $w$ and a colour-seeing Alice in $v$), because the difference-making modal facts aren't in the possible states $w$ and $v$ themselves. Rather, Alice being red-green colourblind in $w$ is the result of the position $w$ has with respect to other epistemically accessible states, and similarly for Alice being colour-seeing. Hence, variation in modal facts is a matter of variation in local structure more than variation in the make-up of the possible states. With statements like (8), the nature of Alice's accessibility becomes even more elusive, for it must encompass not only all those epistemic modal facts that Alice *actually* has but also all the epistemic modal facts that Alice *might* have (for all Bob knows). We can easily imagine how complex the situation can get if we add some further assumptions on Alice's knowledge of Bob's knowledge of her knowledge.

The examples above have illustrated that conventional possible worlds cash out modal facts in terms of the whole structure. In particular, Alice's colourblindness and tree-expertness must be explicated in terms of her total knowledge, instead of the other way around. This holistic nature of the structural component means that possible worlds are not well suited for the task of individuating modal facts. Individuating modal facts is a key feature of any account of possibility that seeks to understand how worlds are built up form their parts. I would also argue that having such a combinatorial account of possibility is the only way to generate a natural semantics for complex higher-order modal statements.

## 4 World Building Revisited

If we want worlds with modal facts as parts, we must start by understanding how to build worlds from parts *simpliciter*.

One of the earliest accounts of world building in terms of atomic parts was given by Wittgenstein in the *Tractatus*. Glossing over the details, we can think of an atomic state of affairs as one or more individuals joined together by a property. The basic world building assumption in the *Tractatus* was that *all* atomic states of affairs are independent of one another, which entailed that any collection of atomic states of affairs could be fused together to make a complex state of affairs. A possible world would then just be something like a maximal complex states of affairs. The problem with this view is that it goes against most of the examples of states of affairs we can muster up. The states of affairs *Entity # 352 is green* and *Entity # 352 is red* appear to be both simple and incompatible. This is basically the problem we encounter with determinate and determinable properties, and some were quick to notice this problem with "tractarian combinatorialism". While the independence of determinable properties seems reasonable enough to grant, the independence of distinct determinates of a same determinable is outright unacceptable. At a minimum, our version of world building will have to take this into account.

### 4.1 Combining Properties and Individuals

The basic suggestion is this. Our primitive building blocks will consist of individuals and properties (both monadic and relational). Given a domain $D$ of individuals and a set of properties $\mathcal{P}$, each one of which has a given arity, a world can be construed as an assignment of extensions to each of these properties. However, as we observed, not any assignment will do because of the exclusive nature of determinate properties of a same determinable. This will constitute our first requirement on assignments: every individual (or tuple of individuals) is assigned at most one determinate property of a same determinable. Call this the exclusiveness condition. Our second requirement on assignments is that every individual (or tuple of individuals) must be assigned at least one determinate property of a determinable. Call this the exhaustiveness condition. Without exclusiveness, we could end up with things that are red (all over) and green (all over). Without exhaustiveness, we could end up with coloured things that have no specific colour.

These conditions can be implemented as follows. We start by requiring that the set of all properties $\mathcal{P}$ be given to us as a partition $\mathbf{P} = \{\mathcal{D}_i : i \in I\}$, where each set $\mathcal{D}_i$ in the partition is meant to represent a collection of mutually exclusive determinate properties of the same determinable type. We will assume that these determinables are pair-wise disjoint, and that all the determinate properties of a same determinable $\mathcal{D}$ all share the same domain $\delta(\mathcal{D})$.[3] A possibility space $\mathcal{A}$ is the structure given by the tuple $\langle D, \mathbf{P}, \delta \rangle$.

An assignment of extensions to properties is a possible world. Each one specifies a maximal combination of individuals and properties. If $w$ is a possible world, the exclusiveness and exhaustiveness mean that: for all determinables $\mathcal{D}$ of $\mathbf{P}$ and for all $\bar{a} \in \delta(\mathcal{D})$, there is a unique $P \in \mathcal{D}$ such that $\bar{a} \in w(P)$. Existence entails exhaustiveness and uniqueness entails

---

[3]The domain restriction is added for extra flexibility and convenience, seeing that some properties aren't obviously applicable to all categories of individuals, e.g., colour and mass to abstract entities.

exclusiveness. The set of possible worlds sp($\mathcal{A}$) generated by $\mathcal{A}$ is simply the set of all assignments $w$ satisfying the exhaustiveness and exclusiveness conditions above. We then say that a property $P \in \mathcal{D}$ holds of $\bar{a} \in \delta(\mathcal{D})$ at $w$ if $\bar{a} \in w(P)$.

A possibility space $\mathcal{B} = \langle D_0, \mathbf{P}_0, \delta_0 \rangle$ is a subspace or substructure of $\mathcal{A}$ iff: (i) $D_0 \subset D$, (ii) $\mathbf{P}_0 \subset \mathbf{P}$ and (iii) $\varnothing \neq \delta_0(\mathcal{D}) = \delta(\mathcal{D}) \cap D_0^k$, for all $\mathcal{D} \in \mathbf{P}_0$ (of arity $k$). Loosely speaking, a subspace is obtained by "forgetting" elements of $D$ and determinables of $\mathbf{P}$ and then restricting the domain of the remaining determinables to the set of remaining elements, but only on the condition that this restriction is non empty. Each world $w_0$ of sp($\mathcal{B}$) is in fact a "part" of a world $w$ in sp($\mathcal{A}$) in the following sense: $w_0$ can be extended to an assignment $w$ of extensions in $D$ to properties in $\mathcal{P}$. In general, there will be more than one way of extending $w_0$. For this reason, if $w_0$ is a world of a subspace, we will often characterize it as a state rather than a world.

## 4.2 Building with Modal Parts

What happens when one of our determinable properties $\mathcal{E} \in \mathbf{P}$ is modal in nature? As it stands, neither the structure nor the world assignments treat $\mathcal{E}$ any differently from a non-modal determinable $\mathcal{D}$. Hence, something must be added to the system to capture the specific behaviour of modal properties.

The characteristic trait of a modal property is that it pertains to other states in the space. The property of being red-green colourblind pertains to other states, namely colour states, in a way properties like red and maple don't. This trait will translate into two added features. First, to each determinable modal property will correspond a subspace of the possibility space, the idea being that the modal property will pertain to this subspace. Second, each determinate modal property will define an accessibility relation (or some structure) on the set of states of the subspace to which it pertains. For example, both red-green colourblindness and colour-seeingness will pertain to the subspace $\mathcal{B}$ generated by the colour determinable, and each will define a (distinct) relation on the set of states of $\mathcal{B}$. This means that we must add to the possibly space functions $m$ and $\rho$ such that: (i) $m$ assigns a subspace $m(\mathcal{E})$ to each modal determinable property $\mathcal{E}$ and (ii) $\rho$ assigns a relation (or structure) $\rho(P)$ on sp($m(\mathcal{E})$) to every property $P \in \mathcal{E}$.

At this point, it may be useful to give an example. Let $\mathcal{A}$ be the possibility space $\langle D, \mathbf{P}, \delta \rangle$. The domain $D$ of this possibility space consists of elements $A$, $a$ and $b$, where $A$ is Alice, $a$ is a tree and $b$ is the collection of leaves in this tree. **P** consists of four determinables $\mathcal{D}_1$, $\mathcal{D}_1$, $\mathcal{E}_1$ and $\mathcal{E}_2$. $\mathcal{D}_1 = \{\text{elm}, \text{map}, \text{oak}\}$ is the tree kind determinable, and $\mathcal{D}_2 = \{\text{re}, \text{gr}, \text{br}\}$ is colour of the leaves determinable. The former has domain $\delta(\mathcal{D}_1) = \{a\}$, and the latter has domain $\delta(\mathcal{D}_2) = \{b\}$. $\mathcal{E}_1 = \{\text{mx}, \text{ox}, \text{tx}\}$ is the "sylvi-epistemic" determinable, and $\mathcal{E}_2 = \{\text{rgc}, \text{cs}\}$ is the "chromo-epistemic" determinable, both of which have domain $\{A\}$.[4] Both are modal determinable properties: $\mathcal{E}_1$ pertains to $\mathcal{B}_1 = \langle \{a\}, \mathcal{D}_1 \rangle$ and $\mathcal{E}_2$ pertains to $\mathcal{B}_2 = \langle \{b\}, \mathcal{D}_2 \rangle$. Each determinate property in $\mathcal{E}_1$ defines an accessibility relation on the set of states $\text{sp}(\mathcal{B}_1)$. There are three states in $\text{sp}(\mathcal{B}_1)$, one corresponding to each tree kind $a$ could be. As for the relations defined: $\rho(\text{mx})$ will be the relation that connects the non-maple states together and the maple state only with itself, $\rho(\text{ox})$ will connect the non-oak states together and the oak state only with itself, and $\rho(\text{tx})$ will connect the states only with themselves. There are also three states in $\text{sp}(\mathcal{B}_2)$, one corresponding to every colour property the leaves could have: $\rho(\text{rgc})$ connects red states to green states and connects the brown state only with itself, and $\rho(\text{cs})$ connects colour states only with themselves.

### 4.3 Aggregating Modal Parts

One may now wonder how these things come together to define Alice's total knowledge, how we can extract from this an accessibility relation on the set of all possibilities that has more or less the same role as the accessibility relation on standard possible worlds.

What we want is to understand Alice's total knowledge as a function of epistemic properties that hold of Alice. The first observation we must make is that what epistemic properties hold of Alice is relative to a world $w$, so Alice's total knowledge will also be relative to $w$. The second observation is that the possibility space approach leaves us with many options as to the way the parts come together to make the whole (which is a good thing). I give here a conservative way of combining modal epistemic parts, one that gives us the standard Kripkean semantic interpretation of an epistemic modality. I would argue, however, that there are alternative ways of combining these epistemic parts that are as interesting as this one.

In order to piece together the modal parts, we must first define the no-

---

[4] mx is maple expertness (can distinguish a maple tree from other trees, but can distinguish other species), ox is oak expertness (can distinguish oaks from other trees, but can't distinguish other species), and tx is total tree expertness (can distinguish all tree species from one another). rgc is red-green colourblindness, and cs no colourblindness at all.

# On Modal Facts in Possible Worlds

tion of the restriction of a world to a subspace. If $\mathcal{B}$ is a subspace of $\mathcal{A}$, observe that every world $w$ of $\mathcal{A}$ can be restricted to a state $w_\mathcal{B}$ of $\text{sp}(\mathcal{B})$ as follows. Since $w$ is a function assigning extensions (of the appropriate arity) to properties in $\mathcal{A}$, when we restrict the domain of $w$ to the properties in $\mathcal{B}$ and intersect each image with $D_0^k$, where $D_0$ is the domain of $\mathcal{B}$ (and $k$ the arity of the property), we obtain a possible state $w_\mathcal{B}$ of $\mathcal{B}$. The relation $R_w$ on $\text{sp}(\mathcal{A})$ that aggregates the epistemic properties Alice has at $w$ is then defined as: $(w, v) \in R_w$ iff, for all epistemic properties $P$ that hold of Alice at $w$, $(w_\mathcal{B}, v_\mathcal{B}) \in \rho(P)$, where $P$ pertains to $\mathcal{B}$. Equivalently, $R_w$ can also be defined as the intersection of all the relations $\rho^*(P)$ on $W$, where $(w, v) \in \rho^*(P)$ iff $(w_\mathcal{B}, v_\mathcal{B}) \in \rho(P)$, with $\mathcal{B}$ being the subspace to which $P$ pertains.

If we look past the mathematical sophistication, the rationale for this definition is quite simple. Suppose $w$ is a world where Alice is a red-green colourblind oak expert, i.e. $A \in w(\text{rgc})$ and $A \in w(\text{ox})$. Since the discrimination powers rgc confers to Alice concern only colour states, it is natural to think that rgc, considered at the level of worlds (and not just colour states), can only discriminate between those parts of worlds that concern colour states, and is completely oblivious to the rest. Similarly, at the level of worlds, ox can only discriminate between those parts of worlds that concern tree states. Aggregating rgc and ox just amounts to intersecting the discrimination powers they have at the level of worlds.

## 4.4 Higher-Order Modal Parts

Defining $\mathcal{A}$'s total epistemic makeup in terms of its parts makes the understanding of higher-order epistemic statements much more transparent. Suppose we want to construct a possibility space that satisfies (8). This could be done by augmenting the example possibility space described above in the following way. We add Bob to the domain of individuals, modify the domains of the existing determinables (to take into account this addition), and add at least one determinable property which qualifies Bob's knowledge of Alice's colourblindness. Bob's knowledge will be a function of the epistemic properties he bears. One of these properties is to be ignorant of Alice's specific colourblindness. Bob's aggregate epistemic accessibility relation will take this fact into consideration very simply, without any convolutions on the part of the accessibility relation.

Suppose we wish to satisfy an even more elaborate modal epistemic statement about Alice:

(9) Alice knows that, for all Bob knows, she might be red-green colourblind or colour-seeing

To satisfy (9), we need only add another epistemic determinable property corresponding to this specific form of knowledge to the structure, one that qualifies Alice's knowledge of Bob's knowledge of Alice's colourblindness. This determinable will pertain to a subspace $\mathcal{E}$, namely the one that qualifies Bob's knowledge of Alice's colourblindness, and its values will determine accessibility relations on the states of $\text{sp}(\mathcal{B})$. Alice's total knowledge is simply obtained by aggregating this relation with the others.

A few remarks before we go on. The examples depicted above all have in common the fact that a modal property never pertains to a subspace of which it is a part. This is not a limitation of the framework. The definition of a possibility space does not forbid modal properties from having this self-pertaining property, nor does it forbid more indirect self-pertaining behaviour: for example, a modal property $\mathcal{E}$ could pertain to a subspace that contains a modal property $\mathcal{E}'$ that pertains to a subspace that has $\mathcal{E}$ as a part. The only intuitive case of a self-pertaining modal property I can think of is awareness. The property of awareness is such that it eliminates states where the agent isn't aware. However, this self-referential aspect of pertaining does not seem to be very widespread.

It should also be clear that the framework above is not limited to representing modal properties as accessibility relations. We can also assign to modal properties things like probability distributions. In fact, when considering the credence state of an agent, this would seem to be the appropriate way to go.

## 5 Propositional Possibility Spaces

If one is interested solely in propositional modal logic, the structures above will seem overcomplicated. In the remainder of this paper, we present a simpler, propositional version of the possibility space framework described above. Instead of properties and individuals (which have no names in a propositional language), we might directly go for states. But if we work directly with states, then we must reformulate our constraints on combinations, since our existing constraints (i.e. exhaustiveness and exclusiveness) are given in terms not of states but of properties and individuals. One approach (cf. Fine, 2011) could be to define a primitive notion of fusion on states and to extract compatibility as a consequence: states are compatible iff they can be fused together. A maximal fusion of states (a fusion of states to which no other state can be added) would then correspond to a possible world. From there, we just modify the framework accordingly. This approach is probably more general, but less user friendly for the purposes we have in mind. Nothing stands in the way of developing the ideas in the way suggested here, but for simplicity we will avoid doing it in that manner.

## 5.1 Possibility Spaces as State Spaces

The simpler version goes like this. We assume there are state "types" and particular states in these states types with the property that no particular state can belong to different state kinds. These particular states are understood to be the "atomic" states in the framework, and particular states of the same kind are incompatible. Let $I$ be the set of all state types and, for $i \in I$, let $\sigma(i) = W_i$ be the set of (atomic) states of type $i$. If $J$ is any subset of $I$, a state of type $J$ is defined as a mapping $w$ from $J$ to states such that $w(i) \in W_i$. We will sometimes call such a mapping a complex or composite state (to contrast them with atomic states). A state of type $J$ is thus an element of the product $W_J = \prod_{i \in J} W_j$. Possible worlds will be states of type $I$, so the set $W$ of all worlds will just be $W_I = \prod_{i \in I} W_i$. Modal states assume more or less the role of modal properties. Hence, modal atomic states of type $i$ will be assigned a set of types $m(i) \subset I$, which consists of those state types to which $i$ pertains. And each particular modal state $s \in W_i$ will determine an accessibility relation $\rho(s)$ on $W_{m(i)}$. In its most general form, a possibility space $\mathcal{S}$ is defined as a triplet $\langle \tau, \sigma, \rho \rangle$, where $\tau = \langle I, M, m \rangle$ is the type of $\mathcal{S}$ ($M \subset I$ is the set of modal state types in $I$).

## 5.2 Syntax and Semantics

A propositional modal language $L$ is given by a set Mod of modalities and a set Prop of propositional variables (or atomic sentences). To interpret $L$ in a possibility space, some additional information is required concerning the modalities. Since modalities are in a sense the syntactic counterparts to modal state types, and since there are typically many modal state types in a possibility space, the model will have to specify what modal states correspond to any given modality. Having said that, we define a model of $L$ as a possibility space $\mathcal{S} = \langle \tau, \sigma, \rho \rangle$, where $\tau = \langle I, M, m \rangle$, together with an interpretation function $val$ such that:

$val(\Diamond) \subset M$, for all $\Diamond \in \mathrm{Mod}$
$val(p) \subset W$, for all $p \in \mathrm{Prop}$

The modal state types in $val(\Diamond)$ are the ones that matter for the evaluation of $\Diamond$.

In order to elegantly present the semantics, we must adopt a few abbreviations. Let $w \in W$ and $J \subset I$. We define $w_J$ as the restriction of $w$ to $J$, i.e. $w_J = w\upharpoonright_J$, and define $w_i$ as the state $w$ assigns to state type $i$. Furthermore, if $v \in W_J$, the point $(w_{-J}, v)$ is the element of $W$ that is identical to $w$ on the state types $I \setminus J$ and identical to $v$ on the state types in $J$.

Given a space $\mathcal{S}$ and interpretation function $val$, truth at a world $w$ of $\mathcal{S}$ is defined as follows. The propositional and Boolean clauses should be straightforward. As for the modal clauses, they will go like this: $w \Vdash \Diamond \phi$ iff, for all $j \in val(\Diamond)$, we have

$$(w_{-m(j)}, v) \Vdash \phi, \text{ for some } v \in W_{m(j)} \text{ such that } \rho(w_j)(w_{m(j)}, v)$$

In other words, $\Diamond$ is interpreted at $w$ with the relation aggregating all the modal states $w_j$, $j \in val(\Diamond)$, where each $w_j$ determines a relation $\rho(w_j)$ on the set $W_{m(j)}$. The meaning of $\Diamond$ at $w$ is therefore determined by the possibility $w$ itself. The definition of validity in a model, in a possibility space and in a collection of possibility spaces follows immediately.

## 6 Concluding Remarks

I have argued here that conventional possible worlds are bad at representing modal facts, and that modal facts are an essential ingredient of a good semantics for modality. This problem was linked to the top-down nature of possible worlds semantics, which made individuation of world parts difficult, if not impossible. An account of world building was given, one that included modal facts as parts, and we saw how it could be put to use to interpret modalities. The framework has shown the potential for promising applications, but I fear that will have to be the object of another paper.

## References

Armstrong, D. A. (1989). *A combinatorial theory of possibility*. Cambridge University Press.

Fine, K. (2011). An abstract characterization of the determinate/determinable distinction. *Philosophical Perspectives, 25*.

Lewis, D. (1986). *On the plurality of worlds*. Blackwell.

Neil Kennedy
FQRSC Post-Doctoral Fellow
MIT, Linguistics and Philosophy
77 Massachusetts Ave.
Cambridge, MA, 2139-4307, USA
e-mail: neil.patrick.kennedy@gmail.com

# Cooperative Question-responses and Question Dependency

### Paweł Łupkowski[1]

**Abstract:** The concept of cooperative question-responses as an extension of cooperative interfaces for databases and information systems is proposed. A procedure to generate question-responses based on question dependency and erotetic search scenarios is presented.

**Keywords:** questions, dependent questions, databases, cooperative answering, inferential erotetic logic, erotetic search scenarios

## Introduction

The main aim of this paper is to propose an extension of cooperative answering techniques with question-responses. In order to obtain such responses I will use the concept of question dependency and erotetic search scenarios—a tool developed within A. Wiśniewski's Inferential Erotetic Logic (IEL).

In the first section I describe the idea of cooperative answering. I also propose to extend standard cooperative answering techniques with a capability of replying with a question. I point out motivations for this step based on the natural language dialogues. In the second section necessary concepts taken from IEL are introduced along with the idea of dependency of questions. Third section contains a description of the procedure which allows to generate cooperative question-replies. The paper ends with indicating potential advantages of supplementing cooperative interfaces with questioning capabilities.

## 1 Cooperative Answers or Cooperative Responses?

The idea behind cooperative answering (in the context of databases and information systems) is to provide a user with an answer to his/her query

---

[1] I would like to give my thanks to D. Leszczyńska-Jasion, M. Urbański and A. Wiśniewski for helpful feedback and comments on a draft of this article. This work was supported by funds of the National Science Council, Poland (DEC-2012/04/A/HS1/00715).

which is not only correct, but also non-misleading and useful (cf. Gaasterland, Godfrey, & Minker, 1992). Let us consider a well known example (cf. Gal, 1988, p. 2) which shall shed some light on what counts as a cooperative answer. Imagine that a student wants to evaluate a course before registering in it. He asks the following question:

Q: *How many students failed course number CS400 last semester?*

Assume also that the course CS400 was not given last semester. For most database interface systems the answer to the student's question would be:

$A_1^*$: *None.*

This answer is correct according to the database information state. But on the other hand, it is easy to notice that it is also misleading for the student (who is not aware of the fact that the course was not given in the last semester) and thus uncooperative from our perspective. However, when we think about a secretary answering the same question we may imagine that the secretary would easily recognise the student's false assumption and correct it in her answer:

$A_1$: *None, but the reason for this is that course number CS400 was not offered last semester.*

The answer $A_1$ is not only correct, but it is also non-misleading and useful for the student. The cooperative answer given by the secretary facilitates student's further search. We may, for example imagine that the next question—asked on the basis $A_1$—would be: *When was CS400 offered and how many students failed it then?*

A review of the literature reveals that techniques developed in the field of cooperative answering are focused on declarative sentences as reactions to the user's queries. In fact, many authors write simply about answers (as declarative sentences). As such, cooperative answering is well explored and a number of techniques have been developed in this area of research, such as: evaluation of presuppositions of a query; detection and correction of misconceptions in a query (other than a false presupposition), formulation of intensional answers or generalisation of queries and of responses. A detailed description of the above techniques may be found in (Gaasterland et al., 1992) and (Godfrey, 1997).

In this paper I will use the distinction—introduced by Webber (1985)—between a *direct answer* and a *cooperative response* to a user's query. A direct answer is described as the information directly requested by the user's query. As for cooperative response, it is described as an informative *reaction* to the user's query (understood here as a declarative sentence or a question). Consequently we may say that the very idea of cooperative interfaces is to provide a user with a cooperative response. The introduced distinction allows us to consider many forms of reactions as responses to the user's query, however two of such reactions are most natural in the context of cooperative answering systems: declarative sentences and questions.

Benamara and Saint-Dizier (2003) gathered a corpus elaborated from Frequently Asked Questions sections of various web services. Besides well formed questions the corpus revealed some interesting types of questions like:

- questions including fuzzy terms (like *a cheap country cottage close to the seaside in Cote d'Azur*);
- incomplete questions (like *What are flights to Toulouse?*);
- questions based on series of examples (like *I am looking for country cottages in mountain similar to Mr. Dupond cottage*).

These questions were not taken into account in WEBCOOP development process (because of early stage of the project at the moment). However, this type of questions asked by users suggests that questions should also be allowed as responses in such systems. Question posing ability would enable to ask a user for missing information not expressed in his/her question in a natural dialogue manner. The motivation comes from everyday natural language dialogues. As Ginzburg points out:

> Any inspection of corpora, nonetheless, reveals the undiscussed fact that many queries are responded with a query. A large proportion of these are clarification requests (...) But in addition to these, there are query responses whose content directly addresses the question posed (...) (Ginzburg, 2010, p. 122)

This fact was also noticed by researchers working with databases. In (Motro, 1994, p. 444) we read:

> When presented with questions, the responses of humans often go beyond simple, direct answers. For example, *a person asked a question may prefer to answer a related question*, or this person may provide additional information that justifies or explains the answer. (emph. by P.Ł.)

## 2  Question Dependency and IEL

As recent corpus study shows, one of the most common question-responses in the natural language conversations are dependent questions (cf. Łupkowski & Ginzburg, 2013). The rationale behind dependent questions might be summarised as follows (Ginzburg, 2010, p. 123): question $Q_1$ depends on question $Q_2$ if discussion of $Q_2$ will necessarily bring about the provision of information about $Q_1$. This allows to say that $Q_2$ might be used to answer $Q_1$—in other words $Q_2$ is an acceptable response to $Q_1$.

The following example illustrates this idea:

> A: Any other questions?
> B: Are you accepting questions on the statement of faith at this point? [F85, 70–71][2]
> (i.e. *Whether more questions exist depends on whether you are accepting questions on the statement of faith at this point.*)

One of the ways in which dependent questions might be modelled is by the use of erotetic implication (e-implication), one of the key concepts of IEL—(cf. Łupkowski, 2012).

In what follows I will use the formal language $L$; this language resembles a language characterised in (Wiśniewski, 2001, pp. 20–21). The 'declarative' part of $L$ is a first-order language with identity and individual constants, but without function symbols. A *sentence* is a declarative well-formed formula (d-wff for short) with no occurrence of a free variable. The vocabulary of the 'erotetic' part of $L$ consists of the signs: ?, {, }, and the comma.

*Questions* of $L$ are expressions of the following form:

$$?\{A_1, A_2, \ldots, A_n\}$$

where $n > 1$ and $A_1, A_2, \ldots, A_n$ are nonequiform, that is, pairwise syntactically distinct, d-wffs of $L$. If $?\{A_1, A_2, \ldots, A_n\}$ is a question, then each of the d-wffs $A_1, A_2, \ldots, A_n$ is a *direct answer* to the question.

A question $?\{A_1, A_2, \ldots, A_n\}$ can be read, 'Is it the case that $A_1$, or is it the case that $A_2$, ..., or is it the case that $A_n$?'.

**Definition 1** *(Wiśniewski, 1995) A question $Q$ implies a question $Q^*$ on the basis of a set of d-wffs $X$ (in symbols:* $\mathsf{Im}(Q, X, Q^*)$*) iff*

---

[2]This notation indicates the British National Corpus file (F85) together with the sentence numbers (70–71).

1. *for each direct answer A to the question Q: $X \cup \{A\}$ entails the disjunction of all the direct answers to the question $Q^*$, and*

2. *for each direct answer B to the question $Q^*$ there exists a non-empty proper subset Y of the set of direct answers to the question Q such that $X \cup \{B\}$ entails the disjunction of all the elements of Y.*

*If $X = \emptyset$, then we say that Q implies $Q^*$ and we write $\mathsf{Im}(Q, Q^*)$.*

The first condition requires that if the implying question is sound[3] and all the declarative premises are true, then the implied question is sound as well[4]. The second condition requires that each answer to the implied question is potentially useful, on the basis of declarative premises, for finding an answer to the implying question. To put it informally: each answer to the implied question $Q^*$, on the basis of $X$, narrows down the set of plausible answers to the implying question $Q$.

Erotetic search scenarios may be defined as sets of the so-called erotetic derivations (Wiśniewski, 2003) or, in a more straightforward way, as finite trees (Wiśniewski, 2010, pp. 27–29), see also (Urbański & Łupkowski, 2010):

**Definition 2** *An e-scenario for a question Q relative to a set of d-wffs X is a finite tree $\Phi$ such that:*

1. *the nodes of $\Phi$ are (occurrences of) questions and d-wffs; they are called e-nodes and d-nodes, respectively;*

2. *Q is the root of $\Phi$;*

3. *each leaf of $\Phi$ is a direct answer to Q;*

4. *$dQ \cap X = \emptyset$;*

5. *each d-node of $\Phi$:*

    *(a) is an element of X, or*

    *(b) is a direct answer to the e-node $Q^*$ which immediately precedes in $\Phi$ the d-node considered (where $Q^* \neq Q$), or*

---

[3] A question Q is *sound* iff it has a true direct answer (with respect to the underlying semantics).

[4] This property may be conceived as an analogue to the truth-preservation property of deductive schemes of inference.

(c) is entailed by (a set of) d-nodes which precede the d-node in $\Phi$;

6. for each e-node $Q^*$ of $\Phi$ different from the root $Q$:

   (a) $dQ^* \neq dQ$ and

   (b) $\mathsf{Im}(Q^{**}, Q^*)$ for some e-node $Q^{**}$ of $\Phi$ which precedes $Q^*$ in $\Phi$, or

   (c) $\mathsf{Im}(Q^{**}, \{A_1, ..., A_n\}, Q^*)$ for some e-node $Q^{**}$ and some d-nodes $A_1, ..., A_n$ of $\Phi$ that precede $Q^*$ in $\Phi$;

7. each d-node has at most one immediate successor;

8. $\Phi$ involves at least one e-node different from the root $Q$;

9. an immediate successor of an e-node different from the root $Q$ is either a direct answer to the e-node, or exactly one e-node;

10. if the immediate successor of an e-node $Q^*$ is not an e-node, then each direct answer to $Q^*$ is an immediate successor of $Q^*$.

The pragmatic intuition behind the e-scenario is that it '... provides us with conditional instructions which tell us what questions should be asked and when they should be asked. Moreover, an e-scenario shows where to go if such-and-such a direct answer to a query appears to be acceptable and goes so with respect to any direct answer to each query.' (Wiśniewski, 2003, p. 422). A certain exemplary e-scenario is presented in Figure 1.

For the proposed approach the idea of a query of an e-scenario is also important.

**Definition 3** A query *of an e-scenario* $\Phi$ *is an e-node* $Q^*$ *of* $\Phi$ *different from the root of* $\Phi$ *and such that the immediate successors of* $Q^*$ *are the direct answers to* $Q^*$.

As it might be noticed e-scenarios are constructed in such a way that all queries are closely related to the initial question by the dependency relation (in IEL expressed in terms of e-implication). We will use this feature in order to generate question-responses. This will ensure that question-responses generated on the basis of e-scenarios will be relevant to the user's question.

Table 1: Example of deductive database

| EDB | IDB | IC |
|---|---|---|
| $usr(a)$ | $locusr(x) \rightarrow usr(x)$ | $\neg(\exists x(live(x,zg) \wedge live(x,p)))$ |
| $usr(b)$ | $locusr(x) \rightarrow live(x,p)$ | |
| $usr(c)$ | $usr(x) \wedge live(x,p) \rightarrow locusr(x)$ | |
| $live(a,p)$ | | |
| $live(b,zg)$ | | |
| $live(c,p)$ | | |

## 3  E-scenarios in Generating Question-responses

In this section I will introduce a simple technique of generating cooperative question-responses on the basis of question dependency check in an e-scenario. Let us assume that we are dealing with a deductive database. It consists of an extensional database (EDB)—built out of facts, intensional database (IDB)—built out of rules, and integrity constraints (IC). I will also assume that there is a cooperative layer between the database and a user where e-scenarios are stored and processed. Let us consider a simple (toy) example of such a database presented in Table 1.

As it might be noticed IDB contains rules for the database. Also new concepts might be introduced here (see the concept *locusr* in Table 1). E-scenarios stored in the cooperative layer are built on the basis of IDB rules (IDB rules are used as premises). For example a relevant e-scenario for a question of the form 'Is $a_i$ a local user?' would fall under the scheme presented in Figure 1.

The following logical facts were used in designing this e-scenario:

1. $\mathsf{Im}(?A, C \rightarrow A, ?\{A, \neg A, C\})$

2. $\mathsf{Im}(?\{A, \neg A, C\}, ?C)$

3. $\mathsf{Im}(?A, B_1 \wedge B_2 \rightarrow A, B_1, A \rightarrow B_2, ?B_2)$

As e-scenarios are stored in the cooperative layer between a user and the database, each question of a user might be processed and analysed against these e-scenarios. I will consider two types of user's questions: (i) about facts (i.e. concerning EDB part) and (ii) about concepts introduced in the IDB part of the database. In both cases the procedure would be the same.

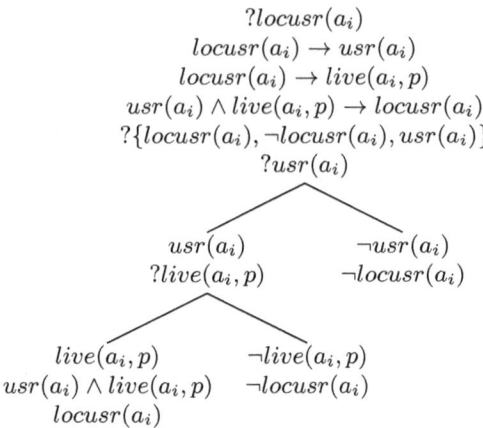

Figure 1: Schema of an e-scenario for a question of the form 'Is $a_i$ a local user?'

The task will be to find user's query among e-scenarios stored in the cooperative layer. When a query is found, its position in the e-scenario should be checked. There are two possibilities:

1. user's question is one of the queries of the e-scenario (in our example, e.g. questions of the form $?live(a_i, p)$);

2. user's question is the initial question of the e-scenario (e.g. questions of the form $?locusr(a_i)$).

Now—on the basis of this search—we may generate two types of question-response before executing user's query against the database:

1. *Were you aware that your question depends on the following questions ...? Would you also like to know their answers?*
   This question-response allows a user to decide how many details he/she wants to obtain in the answer. This also might be potentially useful for future search.

2. *Your question influences a higher level question. Will you elaborate on this subject (follow search in this topic)? May I offer a higher level search?*

# Cooperative Question-responses and Question Dependency

The procedure might be described as follows:
E-scenarios $\{\Phi_1, ..., \Phi_n\}$ ($n \geqslant 1$) are stored in the cooperative layer. By the $uQ$ we designate user's question. $Q_i$ is the initial question of $\Phi_i$ (i.e. the root of $\Phi_i$). $Q_i^*$ is an e-node of $\Phi_i$. The procedure of generating question-responses is the following.

- For each $Q_i$ (where $1 \leqslant i \leqslant n$) check if $uQ = Q_i$.
  - If $uQ = Q_i$, then $uQ$ is a question about a concept introduced in IDB. Return a question-response of the first kind.
  - To report all questions influencing $Q_i$, list all $Q^*$ of $\Phi_i$.[5]
- If $uQ \neq Q_i$, then for each $Q_i^*$ (where $1 \leqslant i \leqslant n$) check if $uQ = Q_i^*$.
  - When you identify $Q_i^* = uQ$, then return a question-response of the second kind.
  - To report higher level question, return $Q_i$ of $\Phi_i$.

Now let us consider some simple examples of questions evaluated against the exemplary database (Table 1).

**Example 1** $uQ_1$: *Is c a local user? (?locusr(c))*.

*On the basis of a schema presented in Figure 1 we generate an e-scenario for question ?locusr(c) by substituting c for $a_i$. Let us refer to this e-scenario as $\Phi_1$. Consequently we will refer to its initial question ?locusr(c) as $Q_1$.*

*In this case $uQ_1 = Q_1$ so the procedure will generate a question-response of the first type. To report all questions on which $uQ_1$ depends the procedure returns all queries of $\Phi_1$, i.e.: ?usr(c) and ?live(c,p). So the response in the natural language form would be:*
*Were you aware that your question depends on the following questions: 'is c a user?' and 'does c live in p?'? Would you also like to know their answers?*

**Example 2** $uQ_1$: *Does c live in p? (?live(c, p))*.

*Also in this case we will use $\Phi_1$. This time $uQ \neq Q_1$, so the procedure tries to match $uQ$ with queries of $\Phi_1$. Such matching is successful for $Q_1^*$: ?live(c, p). The higher level question reported in question-response will be*

---
[5] In the case of really big e-scenarios (with many auxiliary questions) it would be useful to allow a user to decide how many questions influencing $Q_i$ to report.

*simply $Q_1$. So the response in the natural language form would be:*
*Your question influences a higher level question: 'is $c$ a local user?'. Will you elaborate on this subject (follow search in this topic)? May I offer a higher level search?*

What is important, all the necessary data needed to generate question-responses of the analysed kinds might be obtained on the basis of e-scenarios analysis before their execution against the database (i.e. are done in the cooperative layer).[6]

Of course it might be the case that user's question will be identified in more than one e-scenario. Then presented question-responses should report all e-scenarios found. This will have the effect that a user will be aware of contexts in which his/her query is involved in the database. As these are question-responses, the system is expecting user's answer. This answer might be negative, i.e. a user will not use proposed suggestions. In this way the interaction with the system will be closer to the natural language interaction.

## Summary

The proposed simple technique of generating question-replies will enrich cooperative interfaces with question posing capability. What is important, functionality offered by question-replies will be analogical to the functionality of cooperative answers, i.e. it will:

- inform a user (in an indirect manner) about the database schema (this will influence his/her future search and should allow to avoid wrongly formulated questions);

- adjust the level of generality of provided answers to the user's current needs;

- personalise the user's questioning process.

The advantage of question-replies usage will also be a decreased number of database transactions as the question analysis and operations on e-scenarios will be performed in the cooperative layer. As such, question-responses will also supplement the framework proposed in (Łupkowski, 2010).

---

[6] It is worth to mention that also other techniques of cooperative answering might be used with e-scenarios (after their execution)—cf. (Łupkowski, 2010).

Last but not least, involving questions into the cooperative answering process is motivated with natural language dialogues. As a result, interactions with databases and information systems may become more 'natural' and somehow closer to the real-life conversations. Already mentioned corpus study (Łupkowski & Ginzburg, 2013) reveals that in natural language dialogues question-responses addressing the issue of the *way* the answer to the initial question should be given are commonly met. This kind of a question given as an answer might be observed in the following example:

A: Okay then, Hannah, what, what happened in your group?
B: Right, do you want me to go through every point? [K75, 220–221]

As it might be noticed, question-replies proposed in this paper will fit this schema.

## References

Benamara, F., & Saint-Dizier, P. (2003). WEBCOOP: A cooperative question-answering system on the web. In *Proceedings of the tenth conference of european chapter of the Association for Computational Linguistics*.
Gaasterland, T., Godfrey, P., & Minker, J. (1992). An overview of cooperative answering. *Journal of Intelligent Information Systems, 1*, 123–157.
Gal, A. (1988). *Cooperative responses in deductive databases*. Unpublished doctoral dissertation, University of Maryland, Department of Computer Science.
Ginzburg, J. (2010). Relevance for dialogue. In P. Łupkowski & M. Purver (Eds.), *Aspects of semantics and pragmatics of dialogue. SemDial 2010, 14th workshop on the semantics and pragmatics of dialogue* (pp. 121–129). Poznań: Polish Society for Cognitive Science.
Godfrey, P. (1997). Minimization in cooperative response to failing database queries. *International Journal of Cooperative Information Systems, 6*(2), 95–149.
Łupkowski, P. (2010). Cooperative answering and inferential erotetic logic. In P. Łupkowski & M. Purver (Eds.), *Aspects of semantics and pragmatics of dialogue. SemDial 2010, 14th workshop on the semantics*

*and pragmatics of dialogue* (pp. 75–82). Poznań: Polish Society for Cognitive Science.

Łupkowski, P. (2012). Erotetic inferences in natural language dialogues. In *Proceedings of the Logic & Cognition conference* (pp. 39–48). Poznań.

Łupkowski, P., & Ginzburg, J. (2013, March). A corpus-based taxonomy of question responses. In *Proceedings of the 10th international conference on computational semantics (iwcs 2013)* (pp. 354–361). Potsdam, Germany: Association for Computational Linguistics. Retrieved from http://www.aclweb.org/anthology/W13-0209

Motro, A. (1994). Intensional answers to database queries. *IEEE Transactions on Knowledge and Data Engineering, 6*(3), 444–454.

Urbański, M., & Łupkowski, P. (2010). Erotetic search scenarios: Revealing interrogator's hidden agenda. In P. Łupkowski & M. Purver (Eds.), *Aspects of semantics and pragmatics of dialogue. SemDial 2010, 14th workshop on the semantics and pragmatics of dialogue* (pp. 67–74). Poznań: Polish Society for Cognitive Science.

Webber, B. L. (1985). Questions, answers and responses: Interacting with knowledge-base systems. In M. Brodie & J. Mylopoulos (Eds.), *On knowledge base management systems* (pp. 365–401). Springer.

Wiśniewski, A. (1995). *The posing of questions: Logical foundations of erotetic inferences*. Dordrecht, Boston, London: Kluwer AP.

Wiśniewski, A. (2001). Questions and inferences. *Logique et Analyse, 173–175*, 5–43.

Wiśniewski, A. (2003). Erotetic search scenarios. *Synthese, 134*, 389–427.

Wiśniewski, A. (2010). *Erotetic logics. Internal question processing and problem-solving.* (URL=http://www.staff.amu.edu.pl/~p_lup/aw_pliki/Unilog)

Paweł Łupkowski
Chair of Logic and Cognitive Science,
Adam Mickiewicz University
Poznań, Poland
e-mail: Pawel.Lupkowski@amu.edu.pl
URL: http://amu.edu.pl/~p_lup/

# Advice to the Relevantist Policeman

DAVID MAKINSON

**Abstract:** Relevance logic is ordinarily seen as a subsystem of classical logic. At the same time, as has long been known, on the level of formulae of the system it is also a conservative extension of classical logic; but the conventional wisdom has been that the import of this fact is limited as it does not extend to consequence relations. We describe two ways in which it may be thus extended. One is by defining a suitable closure relation out of the set of theses of relevance logic; the other is by adding to the usual natural deduction system for it further rules with 'projective constraints', whose application constrains the subsequent application of other rules. Details and verifications are omitted, but the significance of the two constructions is discussed.

**Keywords:** relevance logic, system R, consequence relations, closure relations, natural deduction, projective constraints

## 1 Introduction

Suppose that, influenced by the persuasion of Anderson, Belnap and a dedicated band of relevance logicians you decide to adopt, not merely as an object of study but in your own reasoning, a system such as the logic R. Are you thereby condemned to forsake parts of classical logic?

The standard response is that you will have to abandon those parts of classical logic that do not appear within the relevance logic. Inference of anything from a contradiction, its dual inferring a tautology from anything, disjunctive syllogism, and various principles for iterated conditionals must all be avoided. Of these, disjunctive syllogism is perhaps the one that most hobbles practical reasoning. Knowing that your favourite socks are in one of the drawers, and having gone through all but the last, you will not be able to make any inference about where they are; you will also have to give up playing Sudoku, which makes incessant use of the principle.

But does acceptance of relevance logic as a tool for use really force us into such restrictions? The system R has as primitive connectives $\wedge, \vee, \neg, \rightarrow$ and the gospel of abnegation presumes that we are systematically replacing the familiar truth-functional connectives by these ones. However, if we think of the arrow as a connective *additional* to the truth-functional ones, there is also a sense in which it has long been known that the system is a conservative extension of classical logic. Anderson and Belnap (1975, section 24.1) already noted the following basic fact: a formula in the language of $\wedge, \vee, \neg$ is a tautology iff it is a thesis of R (indeed, of any of a range of neighbouring systems such as R, E, T, and R-Mingle) in the wider language with arrow added as a connective. Meyer (1974) went further, showing that if we introduce as further primitives a non-classical two-place 'fusion' connective and propositional constants $f, t$, and write $\sim \alpha$ as an abbreviation for $\alpha \to f$, then we can give a straightforward axiomatization of (a conservative extension of) R neatly containing within it a classically complete axiom set for $\wedge, \vee, \neg$ alone.

However, both of these results concern the *sets of formulae* serving as theses of classical and of relevance logic. The accepted wisdom has always been that these results *do not extend to the corresponding consequence relations*; in other words, that classical consequence is not a subrelation of consequence for relevance logic.

But what is the consequence relation for relevance logic? The answer is not cut and dried, even when we have fixed a particular set of formulae as theses, such as those of R. Historically, in both the 1975 and 1992 volumes of their treatise, Anderson & Belnap focussed attention on the set of formulae regarded as acceptable, rather than on a consequence relation between formulae. They did consider notions of derivation-from-assumptions within an axiomatization of the set of acceptable formulae, and also presented systems of natural deduction to generate the same set; but these were not developed into the concept of a consequence relation between a set of formulae on the left and an individual formula on the right, considered in abstraction from any particular path from one to the other. By and large, those working in the wake of Anderson and Belnap have followed suit.

If we forage in the literature then, although there does not seem to be any consequence relation that has been adopted explicitly by relevance logicians, there are eight that naturally suggest themselves. Here upper-case letters are for sets of formulae, lower-case one are for individual formulae.

- One definition would put $A \vdash \beta$ iff for some $\alpha_1, \ldots, \alpha_n \in A$ the formula $(\alpha_1 \wedge \ldots \wedge \alpha_n) \to \beta$ is in R (with this formula read as $\beta$ when $n = 0$).

- A second puts $A \vdash \beta$ iff for some $\alpha_1, \ldots, \alpha_n \in A$ the formula $\alpha_1 \to (\alpha_2 \to \ldots (\alpha_n \to \beta) \ldots)$ is in R. Here the formula embeds arrows.

- The third and fourth are more stringent versions of the above obtained by requiring that all of the formulae in $A$ must appear as antecedents of the certifying formula $(\alpha_1 \wedge \ldots \wedge \alpha_n) \to \beta$ or $\alpha_1 \to (\alpha_2 \to \ldots (\alpha_n \to \beta) \ldots)$ respectively.
- Four more are like the above but also requiring that the premise set $A$ is non-empty.

It must be said immediately that none of these relations extends classical consequence, for example none of them gives us $\{p \wedge \neg p\} \vdash q$. So one may ask: Is there any other way of defining a consequence relation on formulae of relevance logic that leads to R and is nevertheless a conservative extension of classical consequence? In what follows we describe two ways of doing so. Only the main ideas are given: verifications, lesser facts and detailed references to the literature are contained in (Makinson, to appear).

## 2 First Way

We work with relations $\vdash$ between sets of formulae of relevance logic on the left and individual formulae on the right. We do not consider generalizations in which the right position may be occupied by a set of formulae, nor those in which left position is occupied by a multi-set, sequence, or other such structure.

Following familiar terminology, we call $\vdash$ a *closure relation* iff it satisfies the three Tarski conditions of reflexivity ($A \vdash \alpha$ whenever $\alpha \in A$), cumulative transitivity ($A \vdash \gamma$ whenever both $A \vdash \beta$ for all $\beta \in B$ and $A \cup B \vdash \gamma$) and monotony ($A \vdash \alpha$ whenever $B \vdash \alpha$ and $B \subseteq A$). Our question may now be put in a sharper form: is there a closure relation that is a conservative extension of classical consequence to the formulae of relevance logic, and *outputs* R in the sense that for all formulae $\beta$, $\beta \in R$ iff $\emptyset \vdash \beta$?

To construct one, write $\alpha \supset \beta$ as an abbreviation for $\neg \alpha \vee \beta$ and define $A \vdash_0 \beta$ to hold iff for some finite subset $\alpha_1, \ldots, \alpha_n \in A$ the formula $(\alpha_1 \wedge \ldots \wedge \alpha_n) \supset \beta$ is a thesis of R (with this formula understood to be $\beta$ in the limiting case that $n = 0$). Then we have the following: $\vdash_0$ is a closure relation, is compact and closed under substitution, conservatively extends classical consequence, and outputs R. Indeed, it is the least such relation.

The verification is quite straightforward, at various points making essential use of the basic result, recalled above, of Anderson and Belnap (1975, section 24.1), and also the fact that R (likewise E etc.) is closed under detachment for $\supset$, originally established by Meyer and Dunn (1969) and also found in (Anderson & Belnap, 1975, section 25.2.3).

How should we interpret this? On the one hand, it can be seen as providing a motto for the relevantist policeman: *No arrests for classical connections between consenting formulae; they are perfectly legal under a suitably defined consequence relation for relevance logics!*

On the other hand, it could be misleading to describe $\vdash_0$ as itself a *relevant consequence relation*. A minimal condition for such a description to be appropriate for a relation $\vdash$ might be that whenever $\alpha \vdash \beta$ then the formula $\alpha \to \beta$ is in R. Our consequence relation clearly fails that condition. For instance, since it extends classical consequence it tells us that $\alpha \wedge \neg \alpha \vdash_0 \beta$ and $\alpha \vdash_0 \beta \vee \neg \beta$. Also, since $\alpha \supset (\beta \to \beta)$ is shorthand for $\neg \alpha \vee (\beta \to \beta)$, which is a thesis of R, we have $\alpha \vdash_0 \beta \to \beta$. We can, however, describe $\vdash_0$ as a consequence relation *for* relevance logic, in the sense of outputting the theses of the relevance logic R. It conservatively extends classical consequence and, as a bonus, is also a closure relation.

## 3 Second Way

We now consider the question from another angle. As is well known, Anderson & Belnap showed how to generate their system R using a natural deduction system with labels. Can we generate a conservative extension of classical consequence that outputs R in the same kind of way? We show that the answer is positive.

Recall that the labels of the Anderson/Belnap system of natural deduction are supposed to record, for each line in a derivation, the (undischarged) assumptions that were *actually used* to get that line, so that when we come to apply the rule of arrow introduction ($\to+$) we may check the constraint that the supposition being discharged has indeed been used. But, as is notorious, if the rules for conjunction and disjunction are not restricted in some way, they allow 'artificial employment' of assumptions. Quite trivially, with the rules for conjunction, one can apply the $\wedge+$ rule just to get an arbitrary assumption $\beta$ on the payroll and then immediately apply the $\wedge-$ rule to send it on leave. The same may be done using the disjunction rules $\vee+$, $\vee-$ with help from $\to-$: given $\alpha$ we can first use $\vee+$ to infer $\alpha \vee (\beta \to \alpha)$, then carry out two sub-proofs, one obtaining $\alpha$ from the first disjunct simply by reiteration, and the other deriving $\alpha$ from the second disjunct together with the arbitrary assumption $\beta$ by $\to-$, and finally apply $\vee-$ to derive $\alpha$ from $\alpha, \beta$ with both assumptions actually used.

In order to prevent such 'funny business', Anderson & Belnap *restrict* the rules for $\wedge+$ and $\wedge-$ to the special case that the assumptions actually used are the same for the two premises of the rule and, to compensate for the resulting overkill, add to the system a rule expressing distribution of $\wedge$ over $\vee$. At the same time, for reasons quite independent of the 'funny business' problem, the classical rules for negation are truncated, leaving only their de Morgan versions. Not a very elegant dance, as all admit, but it works, at least in the sense of permitting derivation of all and only the theses of the system R from the empty set of assumptions. In particular, the natural deduction system allows derivation of all classical tautologies in $\wedge, \vee, \neg$ while restricting classical consequence, since for example arbitrary formulae are not derivable from explicit contradictions.

The idea behind our natural deduction system is to take exactly the same language, rules, and labelling regime as do Anderson and Belnap and supplement them with rules sufficient for the whole collection to yield classical consequence. However, applications of the additional rules are flagged and act as constraints on subsequent applications of the rule $\to+$ of arrow introduction. The flags are inherited as one passes along a derivation, and arrow introduction is given a second constraint: it cannot be applied when the conclusion of the sub-proof leading up to it is flagged. In this way, the classical strength of the totality of rules is prevented from 'corrupting' the logic of the arrow.

Thus the system monitors the application of all rules by means of both labels and flags, but the flags constrain only the application of $\to+$. We get, one might say, a projective constraint version of the Anderson/Belnap natural deduction system. It provides a second motto for the relevantist policeman: *Take down incriminating classical steps in evidence now, but don't use them until you are facing $\to+$ in the court-room!*

It must be said immediately that projective constraint rules are not new, although they do not appear to have been given a general name nor to have been employed in the context of propositional relevance logic. In classical logic they have for long been used in some natural deduction systems for first-order reasoning. Specifically, some textbook versions of the rule there known as existential instantiation (EI) permit us, under certain conditions, to strip off an existential quantifier and instantiate its free variable, while flagging the step and blocking certain other rules from subsequently being applied to formulae that inherit the flag. Indeed, the Anderson-Belnap labels may also be regarded as 'soft' projective constraints, since they restrict (without banning) subsequent applications of the rules $\to+, \wedge+, \vee+$.

In order to formulate our natural deduction system explicitly, recall that Anderson and Belnap write their labels as subscripts to formulae. But as they also make clear, the same formula may occur at several lines of a derivation with a different status at each line, requiring different labels. Strictly, a derivation in their system should be taken as a finite sequence of line numbers $1, \ldots, n$, with both formulae and labels attached to the line numbers. In turn, labels should be seen as sets of line numbers rather than sets of formulae. Our flag, written as #, will likewise be placed as a superscript to formulae but understood as attached to line numbers. There are two groups of rules.

## Rules of Group I

These are all the rules of Anderson & Belnap for the system R in the connectives $\wedge, \vee, \neg, \to$ (see e.g. Anderson & Belnap, 1975, section 27.2, summarized in Anderson et al., 1992, section R3). The sole difference lies in the formulation of the rule $\to+$, where we impose a new flagging constraint alongside the existing labelling constraint, as italicized in the following.

> $\to+$. Having inferred $\beta_X$ from assumptions $A, \alpha$, we may infer $(\alpha \to \beta)_{X-\{k\}}$ from $A$, where $k$ is the line number of the assumption $\alpha$, provided $k \in X$ and $\beta$ is not flagged.

## Rules of Group II

> *Unrestricted* $\wedge+$. From $\alpha_X, \beta_Y$ infer $(\alpha \wedge \beta)^{\#}_{X \cup Y}$.
>
> *Modus Ponens for* $\supset$. From $\alpha_X, (\alpha \supset \beta)_Y$ infer $\beta^{\#}_{X \cup Y}$.

In both rules the conclusion is flagged. In primitive notation, the second one is a form of disjunctive syllogism, permitting passage from $\alpha_X, (\neg \alpha \vee \beta)_Y$ to $\beta^{\#}_{X \cup Y}$, which is not permitted at all in the Anderson-Belnap natural deduction system, while $\wedge+$ is accepted there in restricted form only: from $\alpha_X, \beta_X$ infer $(\alpha \wedge \beta)_X$.

We also need a general inheritance regime for flags. We need only mention the rules of Group I as those of Group II are automatically flagged.

> *Flagging Inheritance Regime.* In the application of any rule from Group I, if any premise carries a flag then it is inherited by the conclusion.

We say that $\beta$ is a *consequence* of $A$ in our system, and write $A \vdash_1 \beta$, iff there is a derivation in the system with conclusion $\beta$, all of whose undischarged assumptions are in $A$, irrespective of the label or flag that may be attached to $\beta$.

The consequence relation $\vdash_1$ should not be confused with the derivations that generate it. The relation is simply a set of ordered pairs $(A, \beta)$, where $\beta$ is a formula and $A$ is a set of formulae, with no labels, flags, sub-proofs, or intermediate steps, whereas the derivations that generate those pairs may contain all those devices.

Checking the Tarski conditions, reflexivity is immediate. For monotony, the addition of unused premises does not modify the Anderson-Belnap labels, nor add to the flags, so that all steps correct before the addition remain so after it. However, cumulative transitivity fails, as can be illustrated by a simple example. Thus, our consequence relation is not a closure relation. Nevertheless, it does a remarkable job. It is a conservative extension of classical consequence, compact, closed under substitution, and outputs exactly the theses of the relevance logic R.

What is the relationship between $\vdash_1$ and the relation $\vdash_0$ defined in the preceding section? They are presented differently: whereas $\vdash_0$ is defined from the Hilbertian axiomatization of R, $\vdash_1$ is generated by a natural deduction system. They are not the same, since $\vdash_0$ satisfies cumulative transitivity while $\vdash_1$ does not. But there is a tight connection; indeed, it turns out that $\vdash_0$ is just the closure $\vdash_1^+$ of the relation $\vdash_1$ under cumulative transitivity. A word of warning, however: this does not tell us that we get $\vdash_0$ if we add a rule of cumulative transitivity, with some chosen labelling-and-flagging regime, to our projective constraint system of natural deduction. For, if other rules of the natural deduction system are allowed application after applying cumulative transitivity, then $\vdash_1$ could well become a broader consequence relation than $\vdash_0$ and fail to output R.

## 4  General Discussion

The general import of these results depends in part on whether we are thinking in a substantive or a presentational context.

From a *substantive perspective*, they make it clear that if one really wants to be a relevantist, then one may opt for the logic R whilst keeping classical logic intact—rejecting no tautology (as has long been known), impugning no tautological consequence (as shown here). The relevantist enterprise may, after all, be seen as similar to that of conventional modal logic where we add a non-truth-functional connective to a functionally complete set of classical ones and study the outcome; in relevance logic one likewise adds the arrow. This view may even make the relevantist project more attractive than it is at present, since migrating to a non-classical logic is less of a leap into the abyss if we don't have to give up classical logic into the bargain but can draw on it freely within an extended system.

From a *presentational perspective* the picture is rather different. The natural deduction system that we have used to generate the consequence relation $\vdash_1$ is rather ungainly, in that the connectives are poorly sorted out. While the rules of Group II concern only $\wedge, \vee, \neg$ those in Group I (taken from Anderson and Belnap) deal with much more than just $\rightarrow$. Only two of them are 'pure arrow' rules; six are labelled versions of rules for $\wedge, \vee, \neg$; three are 'mixed rules' that combine $\rightarrow$ with $\neg$ or $\vee$. For example, the rule $\neg+$ allows passage from $(\alpha \rightarrow \neg\alpha)_X$ to $(\neg\alpha)_X$, where the arrow figures on the left.

One might expect it to be easy to sort out the connectives more elegantly, but the task is tricky. The present situation thus appears to be as follows. If we wish to use natural deduction to generate a consequence relation for relevance logic, then we can indeed build one, namely $\vdash_1$, that extends classical consequence while outputting the same formulae as R. But the projective constraint system devised for the purpose is a little messier, and considerably more redundant, than the original natural deduction system of Anderson and Belnap whose corresponding consequence relation does not extend classical consequence.

Whether the relevantist project is well-motivated is quite another matter. The author is inclined to suspect that the enterprise is engendered by confusion between two quite distinct problems: that of determining *what is implied* by a given set of assumptions, and that of deciding whether it *would be wiser* to carry out a particular inference in a given situation, or to do something else instead.

The former may be investigated using purely mathematical tools. The latter involves epistemic issues, with questions such as the following. Am I trying to convince my interlocutor (or myself) of the truth of the conclusion of the inference, or merely exploring the more salient consequences of the assumptions, or perhaps constructing a *reductio ad absurdum* refutation of one among them? In the first and second cases, pragmatic considerations also arise. For example, is the conclusion of the inference really worth knowing about and officially registering? Making an inference is like making a journey; we don't go just anywhere our feet can lead us, but travel in an economical or rewarding way to a priority destination. Relevance logic may perhaps be seen as a quixotic attempt to tackle pragmatic issues with purely formal tools.

## Acknowledgements

The author is very much indebted to Diderik Batens, Michael Dunn, Jim Hawthorne, Lloyd Humberstone, João Marcos and Peter Verdée for valuable discussions.

## References

Anderson, A., & Belnap, N. (1975). *Entailment: The logic of relevance and necessity. Vol. I.* Princeton: Princeton University Press.
Anderson, A., Belnap, N., & Dunn, M. (1992). *Entailment: The logic of relevance and necessity. Vol. II.* Princeton: Princeton University Press.
Makinson, D. (to appear). Relevance logic as a conservative extension of classical logic. In S. Hansson (Ed.), *David Makinson on classical methods for non-classical problems.* Berlin: Springer.
Meyer, R. (1974). New axiomatics for relevance logics, I. *Journal of Philosophical Logic, 3,* 53–86.
Meyer, R., & Dunn, J. (1969). E, R, and Gamma. *Journal of Symbolic Logic, 34,* 460–474.

David Makinson
Department of Philosophy, Logic and Scientific Method
London School of Economics
Houghton Street London WC2A 2AE, UK
e-mail: david.makinson@gmail.com
URL: https://sites.google.com/site/davidcmakinson/

# Semantic Facts on Kripke Frames

### JOHANNES MARTI

**Abstract:** This paper addresses the problem of how to represent semantic facts in possible worlds semantics. To this aim we associate a valuation function to every world in a Kripke frame that specifies the language of that world. The result is a two-dimensional semantics for which we present a complete axiomatization in a logical language that is based on standard modal logic. Lastly, we sketch possible philosophical applications of the framework.

**Keywords:** two-dimensional modal logic, meaning change

## 1 Introduction

Many problems that bother philosophers of language crucially concern the relation between semantic notions, such as for instance a sentence expression $p$ being true or $p$ meaning that $\varphi$, and modal or epistemic notions, such as $\varphi$ being necessary or an agent knowing or believing that $\varphi$. Widely known examples are Frege's problem of informative identity statements, the Twin Earth examples and semantic externalism, Kripke's puzzle, the problem of radical interpretation, and the epistemicist's account of vague expressions. Nevertheless, there seems to be no unified logical framework in which the interaction of semantic and epistemic notions can be investigated.

The standard logical framework for modal or epistemic notions is modal logic interpreted with possible worlds semantics. In modal logic semantic questions are typically not considered. The modal formula $\Box p$ is true, if $p$ is true in all the worlds that are accessible from the current world. It is not made explicit whether $p$ is true at a world because of the actual meaning of $p$ and the non-semantic facts of the epistemic alternatives, or because at that epistemic alternative the meaning of the sentence expression $p$ is such that it is true there. Usually it seems to be assumed that semantic facts are fixed and do not vary across worlds.

In (Williamson, 1999) the modal logic B is used to deal with a specific semantic problem. The universal modality $\Box \varphi$ is interpreted as "It is definite that $\varphi$" and used to investigate higher-order vagueness. This interpretation of the modal operator implies that on the semantic side worlds differ only with respect to the semantic facts but not with respect to the atomic facts that language is about. This approach is too restricted to a specific semantic notion to be useful for us.

The formal system that is treated in this paper enriches possible worlds semantics with additional structure to explicitly represents variation in semantic fact. The truth of an atomic expression at a world is made relative to the language in which that expression is evaluated. To access the semantic information contained in valuation models we add a truth operator to the language of modal logic that was already discussed in (Stalnaker, 1978). The resulting system can be seen as a generalization of Stalnaker's metasemantic interpretation of two-dimensional semantics (Stalnaker, 2001, 2004).

The structure of this paper is as follows: First, in section 2, we introduce the semantic structures, we call them valuation models that we use later. In section 3 we discuss different notions of validity in a well-suited modal language. In section 4 the relation to two-dimensional semantics is clarified. In section 5 we sketch illustrative applications of the framework to the problems of radical interpretation and vagueness.

## 2 Valuation Models

The usual semantics for epistemic modal logic is based on Kripke frames (Blackburn, de Rijke, & Venema, 2002). *Kripke frames* are tuples $(W, R)$ where $W$ is any set and $R \subseteq W \times W$ is any relation on the set $W$. The elements of $W$ are called *worlds* and the relation $R \subseteq W \times W$ *accessibility relation*.

In standard Kripke semantics one considers *Kripke models* which are tuples $(W, R, V)$, where $(W, R)$ is a Kripke frame and $V$ a *valuation*, that is function $V : \mathsf{Prop} \to \mathcal{P}W$ for a set $\mathsf{Prop}$ of *propositional letters*. One usually takes the valuation in a Kripke model to fix the basic non-modal facts that are true at the worlds of the model. The atomic fact $p$ is true at all the worlds in the set $V(p)$. In this paper we take a different perspective on valuations. We think of every world in a model as intrinsically containing information on which atomic facts hold at that world. One can then take a valuation as assigning meanings, that is the set of worlds where an expression is true, to atomic sentences. On this view valuations are formal representations of languages. Different valuations correspond to different ways how language might be.

To represent semantic facts in Kripke frames, and hence semantic knowledge for frames under the epistemic interpretation, we exploit the fact that

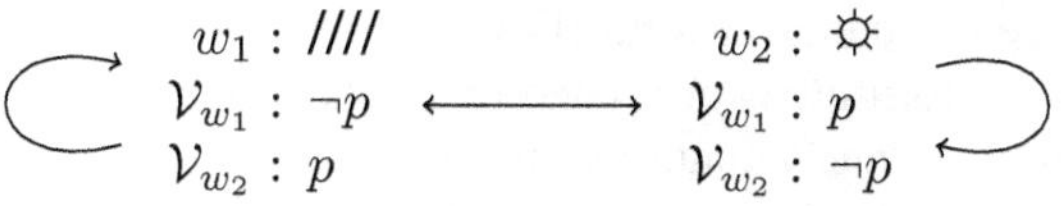

$$w_1 : //// \qquad w_2 : ☼$$
$$\mathcal{V}_{w_1} : \neg p \quad \longleftrightarrow \quad \mathcal{V}_{w_1} : p$$
$$\mathcal{V}_{w_2} : p \qquad\qquad \mathcal{V}_{w_2} : \neg p$$

Figure 1: A Valuation Model

valuations are formal representations of languages. A world in a Kripke frame should not only provide information about its atomic and modal facts but it should also specify the semantic facts that hold there. To do this we associate a different valuation with every world of a frame. Formally, we define a *valuation model* to be a tuple $(W, R, \mathcal{V}_w)_{w \in W}$ where $(W, R)$ is a Kripke frame and $\mathcal{V}_w$ : Prop $\to \mathcal{P}W$ is a valuation for every $w \in W$, that is called the *valuation* or *language of the world* $w$. The valuation $V_s$ represents the semantic facts that hold at the world $w$. In the situation corresponding to the world $w$ an atomic expression $p \in$ Prop has the meaning $\mathcal{V}_w(p)$. The appropriate intuition to have for the languages in a valuation model is that they are different ways how one natural language might be. One should not think of them as being different natural languages such as English or Dutch.

Figure 1 is an example of a valuation model. It contains the two worlds $w_1$ where it is raining, and $w_2$ where the sun is shining. But the worlds do not only differ in weather but also in semantic facts because $\mathcal{V}_{w_1}(p) = \{w_2\}$ but $\mathcal{V}_{w_2}(p) = \{w_1\}$. In the language of $s_1$ the sentence $p$ means that the sun is shining, hence $p$ is true at $w_2$ but not at $w_1$, whereas in the language of $w_2$ the sentence $p$ means that it is raining, hence $p$ is true at $w_1$ but not at $w_2$.

## 3 The Truth Operator

We now define the formal language $\mathcal{L}^T$ which is based on standard modal logic. Aside from the modal box modality it contains an operator T to access the different valuations in a valuation models. The precise syntax of the formal language $\mathcal{L}^T$ is given by the grammar:

$$\varphi ::= p \mid \varphi \wedge \varphi \mid \neg \varphi \mid \Box \varphi \mid \mathrm{T}\varphi,$$

where $p \in$ Prop is any propositional letter.

The intended semantics for the T-operator is to switch the valuation under which its subformula is evaluated to the valuation of the current world. With our intuition that valuations are languages this motivates calling T the *truth operator* for the language of the current world. More precisely the truth conditions of $\mathcal{L}^T$ are as follows: A formula $\varphi \in \mathcal{L}^T$ is *satisfied* at a world $w \in W$ of a model $(W, R, \mathcal{V}_w)_{w \in W}$ under a valuation $V : \mathsf{Prop} \to \mathcal{P}W$ if $w, V \models \varphi$ which is defined inductively as:

$w, V \models p$     iff     $w \in V(p)$,
$w, V \models \varphi \wedge \psi$     iff     $w, V \models \varphi$ and $w, V \models \psi$,
$w, V \models \neg \varphi$     iff     not $w, V \models \varphi$,
$w, V \models \Box \varphi$     iff     $v, V \models \varphi$ for all $v \in W$ with $Rwv$,
$w, V \models T\varphi$     iff     $w, \mathcal{V}_w \models \varphi$.

If $\varphi$ does not contain any propositional letters that are not within the scope of some truth operator its truth value does not depend on the language $V$ and we can just say that $\varphi$ is true at $w$ if $w, V \models \varphi$ for any language $V$.

There are different notions of validity that make sense in the context of valuation models. Here we only consider validity and weak completeness though it should be mentioned all definition and results can be easily adapted to the more general notions of logical consequence and strong completeness.

The most fundamental notion of validity results from varying the world of evaluation and the language of evaluation independently. A formula $\varphi \in \mathcal{L}^T$ is a *general validity* if $w, \mathcal{V}_l \models \varphi$ for all worlds $w$ and $l$ in any valuation model $(W, R, \mathcal{V}_w)_{w \in W}$.

Another notion of validity is obtained by requiring that the language under which a formula is evaluated is the language of the world in which the formula is evaluated. We define a formula $\varphi \in \mathcal{L}^T$ to be *actual-language valid* if $w, \mathcal{V}_w \models \varphi$ for all worlds $w$ in any valuation model $(W, R, \mathcal{V}_w)_{w \in W}$.

An example of an actual-language validity that is not a general validity is the equivalence schema $T\varphi \leftrightarrow \varphi$. It is an advantage of the set of validities as opposed to the set of actual-language validities that they are closed under $\Box$-generalization, that is $\Box \varphi$ is valid whenever $\varphi$ is valid.

Two further notions of logical validity result from either fixing the world of evaluation and quantifying over all languages or to fix the language of evaluation and quantify over all worlds. We define a formula $\varphi$ to be *fixed-world valid* if $w_0, \mathcal{V}_w \models \varphi$ for all worlds $w$ in the valuation model of the world $w_0$. Similarly a formula $\varphi$ to be *fixed-language valid* if $w, \mathcal{V}_{w_0} \models \varphi$ for all worlds $w$ in the valuation model of the world $w_0$. The notions of

# Semantic Facts on Kripke Frames

fixed-world and fixed-language validity are interesting because if we fix the world to be the actual world or we fix the language to be actual English then these notions correspond roughly to the notions of validity that stem from is called interpretational and representational semantics in (Etchemendy, 1990). Special fixed-world validities would include certain peculiar purely modal properties of the actual world. In the context of epistemic logic this might be for instance that a certain agent has inconsistent beliefs. Special fixed-language validities could be certain atomic sentences that are analytic truths such as for instance the English sentence "All bachelors are unmarried".

We only present an axiomatization of general validity since this is much easier than actual language validity or even fixed-world or fixed-language validity. The actual completeness proofs are rather tedious and left out. The set of general validities in $\mathcal{L}^T$ is a normal modal logic in which $\Box$ and T are both normal modal operators. A complete axiomatization with respect to the class of valuation models with arbitrary accessibility relations is given by a normal modal logic K that is additionally closed under generalization for the T-operator and contains the following additional axiom schemata:

$$T(\varphi \wedge \psi) \leftrightarrow T\varphi \wedge T\psi \qquad T\Box T\varphi \leftrightarrow \Box T\varphi$$
$$T\neg\varphi \leftrightarrow \neg T\varphi \qquad TT\varphi \leftrightarrow T\varphi$$

These axioms show how the truth operator distributes over the logical connectives and operators. They correspond to our implicit assumption that the logical symbols have the same meaning in all the languages of a valuation model. For completeness with respect to frames that have a transitive, reflexive relation or an equivalence relation one can just add the S4 respectively S5 axiom schemata for the $\Box$-modality. If one considers transitive, Euclidean frames which are not necessarily reflexive but possibly serial then it is not enough to just add the usual K45 or KD45 axiom schemata for the $\Box$-modality. One can show that in these cases the additional axiom $\Box T(\Box\varphi \to \varphi)$ is needed to obtain completeness.

## 4 Two-Dimensional Semantics

Valuation models can be seen as a formalization of two-dimensional semantics. The two-dimensional framework has been originally introduced to give an account of the meaning for context-dependent expressions such as indexicals and demonstratives (Kaplan, 1978). Later, Stalnaker (1978) used a two-dimensional framework to model how the meaning of an utterance can depend on the facts of the context of utterance. The contextual factors that influence the meaning of an utterance also include the possibility that the meaning of the uttered expressions varies across contexts. With this interpretation of two-dimensionalism, which has been called the "metasemantic" interpretation, there is a close similarity between two-dimensional matrices and valuation models, where the contexts in two-dimensional matrices correspond to the languages in valuation models.

To explain the correspondence between valuation models and two-dimensional matrices we consider again the example in Figure 1. The situation in the valuation model of Figure 1 can also be represented two-dimensionally by the following matrix:

| $p$ | $w_1$ | $w_2$ |
|---|---|---|
| $\mathcal{V}_{w_1}$ | 0 | 1 |
| $\mathcal{V}_{w_2}$ | 1 | 0 |

In this matrix the columns correspond to worlds and the rows to the languages of these worlds. As in two-dimensional semantics the box operator quantifies universally along the horizontal of the matrix, because the original accessibility relation in Figure 1 is a total relation. It is clear that for every valuation model with a total accessibility relation we can find associated matrices for all the propositional letters and whenever we have matrices for all propositional letters over some fixed set of worlds we can find a corresponding valuation model with total accessibility relation. This is the technical sense in which valuation models are a generalization of two-dimensional matrices.

There is also a correspondence between the language $\mathcal{L}^T$ and a logical language suggested in (Stalnaker, 1978). We noticed in the previous paragraph that the $\Box$-modality has the same behavior in both frameworks. In the matrix representation of valuation with total accessibility relation the T-operator shifts the valuation of formulas along the vertical to the diagonal. This is exactly the semantics that Stalnaker specifies for the dagger † in (Stalnaker, 1978). Hence, the two dimensional modal logic of $\Box$ and † is exactly the modal logic with truth operators where the underlying $\Box$-modality is S5. In the case where $\Box$ is a S5 modality $\mathcal{L}^T$ also corresponds to the fragment of the operators ⊟ and ⊕ in the system B of (Segerberg, 1973), which is an early study on two-dimensional modal logic. It shall be pointed out that this logic is different from two-dimensional modal logics containing the $\Box$ modality and an actually operator @ that has been studied extensively for instance in (Blackburn & Marx, 2002; Gregory, 2001; Stephanou, 2005). The actuality operators shifts along the horizontal, the same dimension over which □ quantifies, to the diagonal whereas the truth operator shifts along the vertical.

Given the similarity with two-dimensionalism one can apply the framework of valuation models to the philosophical problems that two-dimensional semantics has been used for. This concerns Frege's problem of informative identity statements, the Twin Earth examples that have been used to argue for externalism of semantic content and Kripke's puzzle about a logically flawless individual that has apparently inconsistent beliefs that it is aware of in different languages. In (Stalnaker, 2004) it is sketched how these puzzles can be resolved with a two-dimensional semantics under the metasemantic interpretation. For a more detailed treatment of these examples in two-dimensional semantics, although not under the metasemantic interpretation, we refer to (Chalmers, 2002).

## 5 Applications

In this section we shortly discuss possible applications of valuation models and of modal logic with truth operators.

### 5.1 Radical Interpretation

In an epistemic logic with truth operators one can study the problem of radical interpretation, as it is discussed in (Davidson, 1984) and more formally in (Lewis, 1974). One might take radical interpretation to be the task of constructing a valuation model to represent the interpreted subject's mental state from the evidence that is available for interpretation. In this section we sketch how this might work in a simple single agent case where it is assumed that the accessibility relation is transitive and Euclidean and hence the logic of □ is K45. We do not discuss any associated philosophical difficulties.

As the basic evidence for interpretation we take the sentences that a subject assents to, or does not assent to, under various circumstances. Such evidence we express with sentences of the form "After the subject has come to believe that $\varphi$ then she does, or does not, assent to $\psi$". Here $\varphi$ is a sentence in the metalanguage of the interpreter that describes some change in the environment that the subject observes, and $\psi$ is a sentence of the subject's own language. To express such sentences in the formal language $\mathcal{L}^T$ we have to find formulas corresponding to the subject assenting to sentence

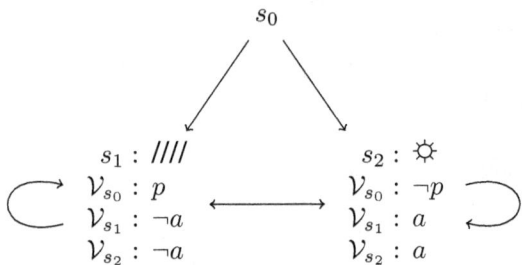

Figure 2: The Subject's Mental State

$\psi$ and the construction that something is true after the subject learned that $\varphi$.

We use the formula $\Box T\psi$ to express formally that the subject assents to the sentence $\psi$. So we require that the subject considers $\psi$ to be true given what she believes about the meaning of the expressions in $\psi$. The formulas of the form $\Box T\psi$ can be seen as a formal representation of Davidson's notion of the subject holding the sentence $\psi$ true in her own language. We take the assent to $\psi$ to express a belief that $T\psi$ which in two-dimensional terminology is called a belief in the diagonal proposition expressed by $\psi$. In (Stalnaker, 1978) it is argued that an assertion of $\varphi$ normally express the belief that $\varphi$, which is called horizontal proposition in two-dimensional terminology. Only if certain pragmatic principles are violated an assertion of $\varphi$ might be reinterpreted to express the diagonal $T\varphi$. In the context of radical interpretation, however, it is justified to take all assertions to express beliefs given by the formula $\Box T\varphi$ instead of just $\Box\varphi$ because the subject's own language might be very different from the interpreters own metalanguage and it is the subject's own language that is of interest.

The whole construction "$\chi$ holds after the subject learned that $\varphi$" can be modeled by formulas of the form $[\varphi]\chi$ where $[\varphi]$ is some sort of update operator from dynamic epistemic logic (Baltag, van Ditmarsch, & Moss, 2008). Here we have not enough space to discuss how update modalities can be made to work in a two-dimensional setting. For our purposes it is enough to consider formulas of the shapes $[\varphi]\Box\chi$ and $[\varphi]\neg\Box\chi$, where $\chi$ does not contain any $\Box$-operators, to be equivalent to $\Box(\varphi \to \chi)$ and $\neg\Box(\varphi \to \chi)$ respectively.

# Semantic Facts on Kripke Frames

Now consider a simple example in which we observe that a subject is assenting to the sentence $a$ if it is not raining and assenting to $\neg a$ if it is raining. Moreover, the subject stays logically consistent so she is not assenting to $\neg a$ if it is not raining and not assenting to $a$ if it is raining. We can summarize this evidence with the following formulas where we assume that $p$ stands for the English metalanguage sentence "It is raining":

$$[p]\, \Box T \neg a \qquad\qquad [\neg p]\, \Box T a$$
$$[p]\, \neg \Box T a \qquad\qquad [\neg p]\, \neg \Box T \neg a.$$

The possible mental states of the subject given the evidence can now be taken to be all valuation models that satisfy the above formulas. One such model is depicted in Figure 2. In this model the language of the actual world $s_0$ is the metalanguage of the interpreter in which the above formulas are evaluated. The propositional letters holding at $s_0$ are not specified since the evidence for interpretation does not constrain the atomic facts at the actual world. One might say that in the model from Figure 2 $a$ means in the language of the subject that it is not raining.

### 5.2 Vagueness

In the definition of valuation models of section 2 languages are identified with valuations that assign a definite truth value, either true or false, to every propositional letters at every world. One might wish to loosen this requirement to accommodate for sentences that do not clearly apply to a situation. An example of this phenomenon are sentences containing vague expressions. In this Subsection I show how the major frameworks that have been suggested to deal with the problem of vague expressions (Williamson, 1994) lead to different ways how one might loosen the requirement that every sentence in a language is either true or false at a world.

One possible adaption to the notion of a language that allows for sentences to be neither true nor false would be to allow the values of propositional letters under a valuation to be at a world to be any element from a set $U$ of generalized truth values. Valuation would then have the type Prop $\to U^W$, that is they are functions that assigns to every propositional letter a function from states to to values in $U$. In the case of standard classical valuations $U$ is just a two element set. A more interesting example would be where $U$ contains contain three values: true, false, and undefined or all real numbers in the interval $[0, 1]$. There are two ways how one could accommodate for many-valued valuations on the syntactic side. One possibility would be to use new operators with the meaning of "$p$ has value $r$" or "the value of $p$ is in $A$" where $r \in U$ and $A \subseteq U$ instead of the usual propositional letters. Alternatively one could also keep the standard propositional letters and propagate the values $U$ along the syntactic tree to all formulas in the formal language $\mathcal{L}^\mathrm{T}$. How this might be done depends on the structure of $U$. In the three valued case one might use the syntactic clauses of K3 or paraconsistent logic and use ideas from (Fitting, 1991) for the modal operator.

Another way to obtain an account of semantic facts that does not require every expression to be either true or false at a world would be supervaluationism. In this case we identify a language of a world not just with one valuation but with a set of valuations. Valuation models are then structures of the form $(W, R, L_w)_{w \in W}$ where $L \subseteq (\mathcal{P}W)^{\mathrm{Prop}}$ is the set of valuation functions that is compatible with the semantic facts holding at the world $W$. The satisfaction conditions of the truth operator could then be changed to:

$$w, V \models \mathrm{T}\varphi \quad \text{iff} \quad w, U \models \varphi \text{ for all } V \in L_w.$$

This definition would give the truth operator the properties of supervaluationistic super truth. It does for instance no longer distribute over negations.

The valuation models from section 2 with classic two-valued valuations allow for an epistemicist's solution to the problem of vagueness if one lets the accessibility relation stand for epistemic uncertainty and the modal $\Box$ for belief or knowledge. With this approach the uncertainty that is involved in the use of vague expression is not at the level of languages but is transferred to the epistemic uncertainty that is modeled by the accessibility relation.

## 6  Conclusions and Future Work

In this paper we present valuation models as a straight-forward extension of Kripke models to represent semantic facts. We worked with a simple modal language $\mathcal{L}^\mathrm{T}$ which has already been suggested in two-dimensional semantics. This language is however not expressive enough for our interpretation of the semantics. For instance one can show that there is no formula in $\mathcal{L}^\mathrm{T}$ that expresses that the propositional letter $p$ means in the language of the current world that $\varphi$. It is an aim for further work to investigate this problem and to extend the formal language.

The relation between valuation models and other interpretations of two-dimensional semantics should be investigated more thoroughly from a philosophical perspective. Originally, two-dimensional semantics has been a tool to capture context dependence and only under the extreme metasemantic interpretation the context can also influence the language in which an expression is uttered. But exactly this extreme case has interested us in this paper. It is an interesting question how the more standard context dependence that arises in the semantics of indexicals fits into the framework of valuation models.

In this paper we suggest applications of valuation models to the problems of radical interpretation and vagueness but we do not discuss them in any detail. We plan to investigate the application to radical interpretation more thoroughly.

A further issue that is not addressed at all in this paper is how different conceptions of what semantic facts are influence the formal model. For instance, if semantic facts are just about the linguistic disposition of agents it might be appropriate to assume epistemic introspection. It might also be interesting to investigate how the choice between internalistic and externalistic conceptions of semantic content influences the framework and how this affects the contents of beliefs that agents can have.

# References

Baltag, A., van Ditmarsch, H., & Moss, L. S. (2008). Epistemic logic and information update. In *Handbook of the philosophy of information*. Elsevier Science Publishers.

Blackburn, P., de Rijke, M., & Venema, Y. (2002). *Modal logic*. Cambridge University Press.

Blackburn, P., & Marx, M. (2002). Remarks on Gregory's "actually" operator. *Journal of Philosophical Logic*, *31*(3), 281–288.

Chalmers, D. J. (2002). The components of content. In *Philosophy of mind: Classical and contemporary readings*. Oxford University Press.

Davidson, D. (1984). *Inquiries into truth and interpretation*. Oxford University Press.

Etchemendy, J. (1990). *The concept of logical consequence*. Harvard University Press.

Fitting, M. C. (1991). Many-valued modal logics. *Fundamenta Informaticae*, *15*, 235–254.
Gregory, D. (2001). Completeness and decidability results for some propositional modal logics containing "actually" operators. *Journal of Philosophical Logic*, *30*(1), 57–78.
Kaplan, D. (1978). On the logic of demonstratives. *Journal of Philosophical Logic*, *8*, 81–98.
Lewis, D. (1974). Radical interpretation. *Synthese*, *27*, 331–344.
Segerberg, K. (1973). Two-dimensional modal logic. *Journal of Philosophical Logic*, *2*(1), 77–96.
Stalnaker, R. C. (1978). Assertion. In P. Cole (Ed.), *Syntax and semantics, volume 9: Pragmatics* (pp. 315–332). New York: Academic Press.
Stalnaker, R. C. (2001). On considering a possible world as actual. *Aristotelian Society Supplementary Volume*, *75*(1), 141–156.
Stalnaker, R. C. (2004). Assertion revisited: On the interpretation of two-dimensional modal semantics. *Philosophical Studies*, *118*(1-2), 299–322.
Stephanou, Y. (2005). First-order modal logic with an "actually" operator. *Notre Dame Journal of Formal Logic*, *46*(4), 381–405.
Williamson, T. (1994). *Vagueness* (Vol. 81). Routledge.
Williamson, T. (1999). On the structure of higher-order vagueness. *Mind*, *108*(429), 127–143.

Johannes Marti
ILLC, University of Amsterdam
P.O. Box 94242
1090 GE Amsterdam, The Netherlands
e-mail: johannes.marti@gmail.com
URL: http://staff.science.uva.nl/~jfmarti/

# The Starring Role of Quantifiers in the History of Formal Semantics

BARBARA H. PARTEE

**Abstract:** The history of formal semantics is a history of evolving ideas about logical form, linguistic form, and the nature of semantics. This paper emphasizes parts of the history of semantics where quantifiers played a major role, including the "Linguistic Wars" of the late 1960's and the conflicts in the philosophy of language between the Ordinary Language philosophers and the Formal Language philosophers. Both conflicts resulted in part from the mismatch between first-order logic and natural language syntax. Both were resolved in part once Montague applied his higher-order typed intensional logic to the analysis of natural language, as illustrated most vividly by the treatment of noun phrases as generalized quantifiers. And quantifiers have played a central role in a number of key subsequent developments in the formal semantics of natural language.

**Keywords:** formal semantics, formal pragmatics, logical form, linguistic form, quantifiers, Montague, Montague grammar, compositionality, history of semantics

## 1 Introduction

There have been centuries of study of logic and of language. Many philosophers and logicians have argued that natural language is logically deficient, or even that "natural language has no logic". And before the birth of formal semantics in the late 1960's, both linguists and philosophers were mostly agreed, for very different reasons, that what logicians meant by "semantics" had no relevance for the study of natural language. But logicians and philosophers of language, even those who regarded natural languages as "illogical" in various ways, made crucial advances in semantic analysis that paved the way for contemporary formal semantics.

The logician and philosopher Richard Montague argued that natural languages do have a very systematic semantic structure, but that it can be understood only if one uses a rich enough logic to mirror the rich type structure that he saw in natural languages. So changing views of the relation between language and logic have often involved changing views of logic itself, and of linguistic structure.

In this paper[1] I'll review some of this background and sketch developments in the growth of formal semantics and formal pragmatics, focusing on crucial turning points in the history of semantics where quantifiers have played a major role. One example: the theory Chomsky described in his 1965 *Aspects of the Theory of Syntax*, where meaning was determined at Deep Structure and transformations were meaning-preserving, ushered in a brief "Garden of Eden" period; what led to expulsion from the Garden and to the Linguistic Wars was (oversimplifying only a bit) linguists' "discovery" of quantifiers. I'll describe this and a number of other crucial points, some earlier and some later. The history of formal semantics is much more than the history of treatments of quantifiers, but their story is an important and fascinating chapter.

## 2 Semantics in Linguistics and the "Discovery" of Quantifiers

Semantics tended to be rather neglected in early and mid-20th century American linguistics[2]. There were several reasons. There had been rather little semantics in early American anthropological linguistics, since in doing linguistic fieldwork one had to start with phonetics, then phonology, then morphology, occasionally a little syntax, and rarely any semantics beyond making dictionaries, or in working out particular lexical domains such as kinship terms. And the behaviorists viewed meaning as an unobservable aspect of language, not fit for scientific study, which influenced the structuralists. And Quine had strong skepticism about the concept of meaning, and had some influence on Chomsky.

---

[1]This paper overlaps with and builds on (Partee, 2011), which provides background for the issues discussed here. Both are part of my work for a book in progress on the history of formal semantics. For the early history of quantification, I have made great use of (Westerståhl, 2011) and (Peters & Westerståhl, 2006), and other sources mentioned in the text. I am grateful for comments and discussion to Paul Pietroski, Alexander Williams, Joshua Stuart Falk, Jon Michael Dunn, participants in Angelika Kratzer's semantics proseminar on quantification in Fall 2011, the participants in the conference "Logica 2012" in Hejnice, Czech Republic, in June 2012, and the audience at my second Baggett Lecture at the University of Maryland in November 2012. All mistakes are my own.

[2]I focus here on linguistics in America, since America is where Montague grammar and formal semantics began. In my book in progress, I will say a bit about the rather different scene in European linguistics.

At the same time there was great progress in semantics in logic and philosophy of language, but that was relatively unknown to most linguists.

In 1954, Yehoshua Bar-Hillel wrote an article in *Language* (Bar-Hillel, 1954) inviting cooperation between linguists and logicians, arguing that advances in both fields made the time ripe for combining forces to work on syntax and semantics together. But Chomsky (1955) rebuffed the invitation, arguing that the artificial languages invented by logicians were too unlike natural languages for the methods the logicians had developed to have any chance of being useful for developing linguistic theory[3].

In *Syntactic Structures* (Chomsky, 1957), Chomsky is quite ambivalent about semantics. He argues that semantic notions are of no use in constructing a grammar, emphasizing that intuitions of grammaticalness are distinct from intuitions of meaningfulness. But at the same time he holds that one test of a good syntax is that it should provide a good basis for a good semantics (if we had any idea how to study semantics). And he argues that transformational grammar is a positive step in that direction, since it uncovers differences at the "transformational level" (what would later be reworked as "deep structure") that are obscured in the output (later "surface structure").

But Chomsky also notes that transformations sometimes change meaning. "... we can describe circumstances in which a 'quantificational' sentence such as (1a) may be true, while the corresponding passive (1b) is false, under the normal interpretation of these sentences—e.g., if one person in the room knows only French and German, and another only Spanish and Italian. This indicates that not even the weakest semantic relation (factual equivalence) holds in general between active and passive." (pp. 100–101)

(1) a. Everyone in this room knows at least two languages.

b. At least two languages are known by everyone in this room.

In later years, those judgments about (1) came to be questioned; some argued that (1b) is ambiguous, some argued that both are. Chomsky himself noted problems with the judgments and their diagnosis in (Chomsky, 1965). Difficulties with such data continued for many years. Over time, linguists have developed more subtle ways to get data than just asking about their own or their consultants' intuitions. But the unclear relation between En-

---

[3]When David Kaplan heard about Chomsky's reply to Bar-Hillel, he said it reminded him of Quine's vehement rejection of Kripke's work on modal logic (David Kaplan, p.c. January 2011).

glish (and not only English) syntax and quantifier scope has continued to be a topic of much study, many proposals[4], and little consensus.

**Early Semantics in Generative Grammar**

Katz and Fodor (1963) added a semantic component to Chomsky's generative grammar. They addressed the "Projection Problem", i.e. compositionality: how to get the meaning of a sentence from meanings of its parts. At that time, "Negation" and "Question Formation" were transformations of declaratives: they were prime examples of meaning-changing transformations. So at that time, it was accepted that meaning depended on the entire transformational history.

In a theoretically important move, related to the problem of compositionality, Katz and Postal (1964) made the innovation of putting such morphemes as Neg(ation) into the Deep Structure, as in (2), arguing that there was independent syntactic motivation for doing so; they hypothesized that transformations never change meaning, and that meaning could be determined on the basis of Deep Structure alone. The revised Negation transformation T-NEG did not change meaning; its job was to put the negative morpheme into its appropriate form and position in the Surface Structure, the level that provides the input to phonological rules that determine how the sentence is pronounced.

(2) [Neg [Mary [has [visited Moscow]]]] $\Rightarrow_{\text{T-NEG}}$
   Mary [has not [visited Moscow]]

In Aspects of the Theory of Syntax (Chomsky, 1965), Chomsky tentatively accepted Katz and Postal's hypothesis. The architecture of the theory—syntax in the middle, mediating between semantics on one side and phonology on the other—was elegant and attractive. This big change in architecture rested on the claim that transformations should be meaning-preserving. This led to what I call the "Garden of Eden" period around 1965, when there was widespread optimism about the Katz-Postal hypothesis, and the syntax-semantics interface was believed to be relatively straightforward (even without having any really good ideas about the nature of semantics.)

What happened to upset that lovely view? Although of course there were multiple factors, I think it's fair to focus on one salient issue: linguists

---

[4]Quantifier scope ambiguity in natural language remains a big and important topic; I will neglect it here for reasons of space, but in the planned book I will discuss at least 10 proposals for analyzing it.

discovered quantifiers (Bach, 1968; Karttunen, 1968, 1969; Lakoff, 1968; McCawley, 1971). Transformations that preserved meaning (more or less) when applied to names clearly did not when applied to some quantifiers. Clear examples come from "Equi-NP Deletion" (Rosenbaum, 1967), the transformation that applied to (the structure underlying) (3a) to give (3b).

(3) a. John wants John to win.

b. John wants to win.

When the identical NPs are names, the transformation preserves meaning. But applied to sentences with quantifiers, it has the unwanted result of deriving (4b) from (4a).

(4) a. Everyone wants everyone to win.

b. Everyone wants to win.

It was a surprising historical accident that the behavior of quantifiers was not really noticed until the Katz-Postal hypothesis had for most linguists reached the status of a necessary condition on writing rules. Here are a few more examples of derivations that would have been given in Chomsky's 1965 "standard" theory; I doubt that the Katz-Postal hypothesis would have been suggested if these had been noticed earlier.

(5) a. Every man voted for himself. <u>FROM</u>:

b. Every man voted for every man.

(6) a. All pacifists who fight are inconsistent. <u>FROM</u>:

b. All pacifists fight. All pacifists are inconsistent.

(7) a. No number is both even and odd. <u>FROM</u>:

b. No number is even and no number is odd.

The problems illustrated by examples (4-7) showed the untenability of the Katz-Postal hypothesis combined with the *Aspects* theory of syntax, and led to expulsion from the Garden of Eden and to the "Linguistic Wars" (Harris, 1993; Newmeyer, 1980) between the Generative Semanticists and the Interpretive Semanticists. Generative Semanticists held onto the goal of compositionality and pushed the 'deep' structure deeper, making it a kind of "logical form". Chomsky had been tentative about adopting the Katz-Postal hypothesis in the first place, and valuing syntactic autonomy more highly, abandoned it. The linguistic wars raged from the late 1960's into the mid 1970's.

What were the early linguistic notions of 'logical form'? Generative Semanticists (Lakoff, Ross, McCawley, Postal, and others) (see Newmeyer, 1996), argued that in order for deep structure to capture semantics, it needed to be deeper, more abstract, more like "logical form", which for linguists meant first-order-logic. In fact, both linguists and philosophers who worried about the semantics of quantified sentences before Montague's work thought of "logical form" in terms of first-order logic. But given that generalized quantifiers were only developed starting in the late 1950's, that could hardly have been otherwise. I return in Section 4 to the role of generalized quantifiers in reconciling the apparent mismatches between "linguistic form" and "logical form" of sentences with quantifiers. But first we need a quick review of some of the crucial developments from Aristotle to Frege and Tarski.

## 3 Developments in Logic

This section is short, since most of this material is familiar to the Logica readership[5]. I will just review some relevant history of quantifiers that forms part of the background for the development of formal semantics.

When Aristotle (384-322 B.C.E.) invented logic, he focused on quantification; operators like *and* and *or* were added by the Stoics. Implicit in Aristotle's syllogistic is a semantics for the quantifiers. Each of the four quantifier expressions can be seen as standing for a binary relation between properties:

(8) (i) *all* $(A, B) \Leftrightarrow A \subseteq B$

(ii) *some* $(A, B) \Leftrightarrow A \cap B \neq \emptyset$.

etc.

In hindsight, this is close to the idea of Generalized Quantifiers. But the idea of giving a semantic value to the quantifiers themselves was not explicitly

---

[5]For help with this section I am indebted to more conversations and sources than I can remember; my sources include at least (Cocchiarella, 1997; Stanley, 2008) and conversations with Dagfinn Føllesdal, Joseph Almog, David Kaplan, and others, in addition to my own rather eclectic education in philosophy.

developed until much later, really not until Frege. In the Middle Ages there was a lot of work trying to figure out meanings for expressions like *all men* or *some man*, sometimes syncategorematically (Albert of Saxony analyzed *all* and *some* via conjunctions and disjunctions), sometimes categorematically (giving them independent meanings) with convoluted theories.

Leibniz may have been the first to use bound variables, but it was in his integral calculus, not in logic. Those variables were intrinsically bound, not replaceable by constants. A different use of variables was already in use in algebra, in formulas like (9).

(9) $x + (y + z) = (x + y) + z$ $\qquad$ Law of Associativity

These variables could be substituted for by constants; they were implicitly bound by universal quantifiers.

But those two uses were not united and generalized until Frege.

The word 'quantifier' appears first[6] in (De Morgan, 1862), as an abbreviation for Hamilton's 'quantifying phrase'. The American philosopher and logician C.S. Peirce (1839-1914) is often credited with developing the theory of relations (Burris, 2009).

### Frege

The greatest foundational figure for formal semantics is Gottlob Frege (1848-1925). His crucial ideas include the Compositionality Principle and the idea that function-argument structure is the key to semantic compositionality.

*The Principle of Compositionality:* The meaning of a complex expression is a function of the meanings of its parts and of the way they are syntactically combined.

One of Frege's great contributions was the logical structure of quantified sentences. That was part of the design of a "concept-script" (*Begriffschrift*), a "logically perfect language" to satisfy Leibniz's goals; he did not see himself as offering an analysis of natural language, but a tool to augment it, as the microscope augments the eye.

### Does Ordinary Language 'Have no Logic'?

As Russell, Carnap, and Tarski were making advances in logic and the philosophy of language, a war began within philosophy of language, the "Or-

---
[6]Source: Wilfrid Hodges' web page: http://wilfridhodges.co.uk/.

dinary Language" vs "Formal Language" war. Ordinary Language Philosophers rejected the formal approach, and urged more attention to ordinary language and its uses. Strawson said in 'On referring' (Strawson, 1950): "The actual unique reference made, if any, is a matter of the particular use in the particular context; ... Neither Aristotelian nor Russellian rules give the exact logic of any expression of ordinary language; for ordinary language has no exact logic." Russell (1957) replied, "I may say, to begin with, that I am totally unable to see any validity whatever in any of Mr. Strawson's arguments." But near the end of the paper, he adds, "I agree, however, with Mr. Strawson's statement that ordinary language has no logic."

Russell was not the first logician to complain about the illogicality of natural language. One of his favorite complaints was that English puts phrases like "every man", "a horse", "the king" into the same syntactic category as proper names. He considered the formulas of his first-order logic a much truer picture of 'logical form' than English sentences.

An exercise I often give my students to help them appreciate Montague's use of a higher-order typed logic, including generalized quantifiers, is to consider the question of where in Russell's formula (10), symbolizing *Every man walks*, is the meaning of *every man*?

(10) $\forall x(\text{man}(x) \rightarrow \text{walk}(x))$

The answer is that it is distributed over the whole formula—in fact everything except the predicate *walk* in the formula can be traced back to *every man*. One way to answer Russell is to devise a logic in which the translation of *every man* is a constituent in the logical language. Terry Parsons did it with a variable-free combinatory logic (Parsons, 1968, 1972). Montague did it with a higher-order typed intensional logic (Montague, 1973). Both were reportedly[7] influenced by seeing how to devise algorithms for mapping from (parts of) English onto formulas of first-order logic, thereby realizing that English itself was not so logically unruly. (See also D. Lewis, 1970.) First-order logic has many virtues, but similarity to natural language syntax is not one of them.

**More on Frege and Tarski**

Frege worked out the semantics of free and bound variables, and developed the syntax and semantics of quantifiers as variable-binding operators.

---

[7] Terry Parsons, p.c. In the case of Montague, a number of people have pointed to the attention to translation between English and logic in (Kalish & Montague, 1964).

And in a sense he did it more compositionally than Tarski. In Tarski's semantics for quantified sentences, standard in logic textbooks, the quantifier symbols ∀ and ∃ are not themselves given a semantic interpretation. They are treated *syncategorematically*: we are given semantic interpretation rules for formulas *containing* quantifiers. Tarski's semantics is thus not strictly compositional.[8] Tarski does not get the interpretation of (10) by combining the interpretation of the quantifier with the interpretation of the rest of the formula; instead he has a schema that gives the interpretation of (10) by considering satisfaction of the open formula by all possible values of the variable $x$.

Frege treated the quantifier symbols as categorematic, standing for certain second-order objects (Peters & Westerståhl, 2006, pp. 35–40). Although his notation was quite different from modern notation, he treated the universal quantifier as a unary second-level operator that applies to a first-level predicate to give a truth-value. Peters and Westerståhl observe that Frege thus invented a kind of generalized quantifier, though it was forgotten until reinvented in a model-theoretic context.

Frege's universal quantifier was "everything" rather than "every". We can represent it (not in Frege's own notation) as a set of sets as in (11) (where $D$ is the universe):

(11) $\lambda P.\forall x P(x)$ or $\{P : D \subseteq P\}$.

A sentence like *Every boy walks* would be paraphrased into something like "Everything is such that if it is a boy, it walks." Thus Frege's analysis of universal quantifiers had some things in common with later generalized quantifiers, but like Russell, he did not directly analyze NP constituents like *every boy*.

Tarski (1902-1983) developed model theory based in set theory and with it made major advances in providing a semantics for logical languages. Frege had had an absolute notion of truth, and a single domain of all objects; all non-variables had fixed interpretations. It has been common to trace the key notion of model theory, *satisfiability-in-a-structure*, or *truth-in-a-model*, back to Tarski's seminal paper (Tarski, 1935) on the concept of truth in formalized languages. Hodges (1985/6) argues that it was only in the 1950's that Tarski introduced interpretation relative to models. But Hodges'

---

[8]This has been reported in various works; I learned it, with some embarrassment, from Tarski (p.c.) after having said, in a talk about Montague Grammar for the Berkeley logic group in the 1970's where he was present, that the compositionality of Tarski's semantics for first-order logic was a model for Montague's work on natural language.

arguments are disputed by Niiniluoto (1994) and by Feferman (2004), who finds the notion of *truth-in-a-structure* present implicitly in Tarski's work as early as 1931.

In any case, model theory revolutionized semantics. This comes out most clearly (for linguists) when we look at the cascade of advances that came with the study of generalized quantifiers, a few of which will be described in Section 4. There had been work on generalized quantifiers by Mostowski (1957) and Lindström (1966) before Montague's work, but for formal semanticists the source of generalized quantifiers was Montague and David Lewis[9].

## 4 Generalized Quantifiers

### Montague's Work on Quantifiers: Background

Montague, a student of Tarski's, contributed greatly to the development of formal semantics with his development of intensional logic and his combination of pragmatics with intensional logic (Montague, 1968, 1970). His higher order typed intensional logic unified modal logic, tense logic, and the logic of the propositional attitudes, extending the work of Carnap (1956), Church (1951), and Kaplan (1964), putting together Frege's function-argument structure with the treatment of intensions as functions to extensions.

Montague treated both worlds and times as components of "indices", and intensions as functions from indices (not just possible worlds) to extensions. The strategy of "add more indices" was taken from Dana Scott's "Advice on modal logic" (Scott, 1970), an underground classic long before it was published.

The Fregean principle of compositionality was central to Montague's theory and remains central in formal semantics. Montague showed that one could give a model-theoretic semantics for ordinary English, with a syntax rather close to surface structure; his higher-order typed logic[10] was crucial

---

[9] David Lewis presented the idea in a talk in 1969 and Montague in 1970; see (Lewis, 1970) and (Montague, 1973). They were colleagues, and no one seems to know whether they developed the idea independently, or together, or whether Montague got the idea from Lewis. Probably because Montague's semantic program was more comprehensive, his work on this topic is cited more often.

[10] Montague's views of natural language type structure had a number of influences, including categorial grammar as developed by Polish logicians; he cited (Ajdukiewicz, 1960).

for making that possible. The treatment of English noun phrases as uniformly denoting generalized quantifiers was one of the most vivid examples of that; that achievement made a big impression on linguists[11].

**Montague and Generalized Quantifiers**

According to Peters and Westerståhl (2006), the logical notion of quantifiers as second-order relations is "discernible" in Aristotle, full-fledged in Frege, then forgotten until rediscovered by model theorists. Mostowski (1957) introduced unary generalized quantifiers, denoting sets of sets; these can capture the meanings of quantified expressions like *everything, something, an infinite number of things, most things*. It was Lindström (1966) who introduced binary generalized quantifiers, without which one can express *Most things walk*, but not *Most cats walk*. What we are now accustomed to calling 'generalized quantifiers', e.g. the denotation of the NP[12] *most cats*, represents the application of a Lindström binary quantifier *most*, syntactically a Determiner, to its first argument *cats*, syntactically a Noun (or Common Noun Phrase), giving a unary generalized quantifier *most cats*.

Montague (1973) (and D. Lewis, 1970) proposed that English NPs like *every man, most cats* can be treated categorematically, uniformly, and compositionally if they are interpreted as generalized quantifiers. This was a big part of the refutation of the point Russell and Strawson (and Chomsky) were agreed on, that there is no logic of natural language. That refutation opened a floodgate, and the next decade saw a great surge of work by linguists and philosophers, individually and together.

**Generalized Quantifiers and English Syntax**

Montague's work showed how with a higher-typed logic and the lambda-calculus (or other ways to talk about functions), NPs could in principle be uniformly interpreted as generalized quantifiers (sets of sets). And Determiners could then be interpreted uniformly as functions that apply to common noun phrase meanings (sets) to make generalized quantifiers.

---

[11] In my own case, it was one of the things that made me devote a good number of years to finding ways to integrate Montague's work into linguistics. The enterprise started out as "Montague Grammar" (Partee, 1973, 1975, 1976), and developed into the more inclusive field of formal semantics.

[12] In much western syntax, the Determiner is now considered the head of the Noun Phrase (Abney, 1987), which has been rechristened Determiner Phrase (DP). In this paper I continue to call it an NP.

Recall how we asked "Where's the meaning of *every man* in (10), the first-order formalization of *Every man walks*?" Now, with Montague's work, we have a semantic type, $\langle\langle e,t\rangle,t\rangle$, sets of sets of entities, to correspond to English NPs. In a simple sentence, the subject NP is the function, and the Verb Phrase (VP) is its $\langle e,t\rangle$-type argument, as shown in (12). And although Montague treated the determiner *every* syncategorematically, that was inessential; *every* can be analyzed as in (12d).

(12) a. *Every student*
$\lambda P \forall x[\textbf{student}(x) \to P(x)]$      type $\langle\langle e,t\rangle,t\rangle$
b. *walks*
**walk**      type $\langle e,t\rangle$
c. *Every student walks*
$\lambda P \forall x[\textbf{student}(x) \to P(x)](\textbf{walk})$      type $t$
$\equiv \forall x[\textbf{student}(x) \to \textbf{walk}(x)]$
d. *every*
$\lambda Q \lambda P \forall x[Q(x) \to P(x)]$      $\langle\langle e,t\rangle,\langle\langle e,t\rangle,t\rangle\rangle$

Montague's interpretation of the sentence *Every man walks* is the same as Russell's; the big difference is that Montague derives the interpretation compositionally; the semantic structure is homomorphic to the syntactic structure.

Stokhof (2006), in describing the PTQ model of Montague grammar, isolates "two core principles that are responsible for its remarkable and lasting influence":

A. Semantics is syntax-driven, syntax is semantically motivated.

B. Semantics is model-theoretic.

Montague did not invent model-theoretic semantics; but it was through his work that the model-theoretic approach became more widely known and adopted among linguists, with far-reaching changes to the field of linguistic semantics.

## Generalized Quantifier Theory and Model Theory

Barwise and Cooper, a logician and a linguist, cooperated in the first major investigation of properties of determiners (Barwise & Cooper, 1981), studying English noun phrases from the perspective of the model-theoretic properties of generalized quantifiers and the determiners that help to build them.

A determiner like *every* denotes a function of type $\langle\langle e,t\rangle,\langle\langle e,t\rangle,t\rangle\rangle$ as seen in (12d) above. The determiner's first argument is the common noun: [[*every*]]([[*student*]]). This generalized quantifier is itself of type $\langle\langle e,t\rangle,t\rangle$, and takes the denotation of the VP as its argument: [[*every*]]([[*student*]])([[*walks*]]). Schematically the structure of a simple sentence of that sort can be represented as $D(A)(B)$.

One of the first determiner universals discovered by Barwise and Cooper is the "Conservativity" universal expressed in (13):

(13) **Conservativity Universal**: Natural language determiners denote conservative functions.

(14) **Definition**: A determiner meaning $D$ is conservative iff for all $A, B, D(A)(B) = D(A)(A \cap B)$.

(15) **Examples**:

No solution is perfect = No solution is a perfect solution.

Exactly three cups are blue = Exactly three cups are blue cups.

Every boy is singing = Every boy is a boy who is singing.

For example, no language has a (non-conservative) determiner $D$ such that *D members are excluded* would mean *All non-members are excluded*. And if the word *only* in *only boys* were a determiner, it would be an example of a non-conservative determiner, since the false sentence *Only males are astronauts* is not equivalent to the true sentence *Only males are male astronauts*. But *only* is not a determiner; in general it combines with an expression of category X to make a new expression of category X, and in examples where it seems to combine with a noun to make an NP, it is really combining with an NP to make another NP, as it does in *only that boy, only John, only John and the teacher*.

The Conservativity Universal suggests a reason why it is useful for languages to have determiners that combine with a noun to make a generalized quantifier. The Conservativity Universal tells us that when evaluating $D(A)(B)$, we only need to consider $A$'s, never non-$A$'s. And the "$A$" position in that formula corresponds to the noun *student* in *every student walks*. So the noun indicates the domain of entities that are relevant to the truth of any sentence of the form D NP VP. So not only are natural languages not "illogical"; the more we uncover about how the compositional semantics works, the more well-designed (well-evolved) natural languages turn out to be[13].

Barwise and Cooper had many other results of interest to linguists and logicians. They found a first good approximation to a formalization of the distinction between "weak" determiners, which can occur in *there*-sentences like (16a), and "strong" determiners, which cannot (16b)[14].

(16) a. There are some/three/several/many/no unicorns in the garden.

b. *There are both/the/those/all unicorns in the garden.

(17) Key definitions:

a. A determiner $D$ is *positive strong* if $D(A)(A)$ is true whenever $D(A)$ is defined ($A$ any subset of the universe).

b. $D$ is *negative strong* if $D(A)(A)$ is false whenever $D(A)$ is defined.

c. $D$ is weak if it is neither positive strong nor negative strong.

For a fuller discussion, and many more interesting properties of determiners and generalized quantifiers, see (Keenan, 2003; Keenan & Westerståhl, 1997; Larson, 1995; Westerståhl, 2011).

## 5 Quantifiers and Pragmatics

Quantifiers also played an important role in the development of formal pragmatics and the recognition of the necessarily close connection between formal semantics and formal pragmatics.

One interesting domain where this can be seen is in the history of work on "Negative polarity items" or NPI's. These are linguistic expressions like *ever* and *at all* and some uses of *any*, which sound fine in negative sentences but not in simple positive sentences.

(18) a. John doesn't *ever* eat *any* vegetables *at all*.

b. *John *ever* eats *any* vegetables *at all*.

---

[13] I am not claiming that natural languages are the best medium for "doing logic"; I subscribe to Frege's statement that a formal language provides an aid to natural language the way a microscope aids the eye.

[14] An asterisk in front of a sentence indicates that the sentence is ill-formed.

A "negative" determiner like *no* in subject NP licenses NPIs as in (19a), but a "positive" determiner like *some* does not, as seen in (19b).

(19) a. No boy had *ever* seen *any* problem *at all* in *any* of her proposals.

b. *Some boy had *ever* seen *any* problem *at all* in *any* of her proposals.

For a long time most linguists had thought that what was crucial for allowing NPIs in a sentence was the presence of some "negative" word or morpheme in a suitable structural position—hence the name "Negative polarity items". But it was known that there were contexts without any overtly negative elements that nevertheless allowed NPIs: comparative clauses, as in (20a); the antecedent but not the consequent of a conditional, as in (20b,c), and in the first argument but not the second argument of the determiner *every* as in (20d,e).

(20) a. Mary answered more questions than *anyone* had *ever* answered before.

b. If you *ever* have *any* problem *at all*, please let me know.

c. *If John is lazy, he will have *any* problems *at all*.

d. Every child who *ever* went there received some gifts.

e. *Every child who went there at least once received *any* gifts.

That last pair, (20d,e), came as a surprise when it was discovered by Ladusaw (1979), since previously linguists had thought that determiners could be divided into negative ones like *no* that do allow NPIs and positive ones like *some* that do not. Ladusaw identified the crucial model-theoretic generalization stated in (21).

(21) **NPI Generalization (Ladusaw)**: NPIs can occur in expressions that form part of the argument of a monotone decreasing function.

Ladusaw's generalization was the first example of a linguistic puzzle solved by model-theoretic means that could not be mimicked by some more "syntactic" tree-like representation at some level of "logical form", the way one can syntactically show scope relations and function-argument structure. The crucial property of being a monotone decreasing function has no "logical form" representation; it is an inherently model-theoretic property.

Ladusaw's generalization was a major milestone in formal semantics, and at the same time opened up two important problems for further work. (i) He gave no account of the meanings of the NPIs themselves, treating them as having simple existential meanings. The meaning of *any* in the examples in (20) is treated as basically the same as the meaning of *a* or *some*, restricted to narrow scope under its "licensing" function expression. (ii) The licensing of NPIs is sometimes sensitive to inferential properties in context, and not only to properties of the semantic content of the sentence. Examples like (22a-b), from (Ladusaw, 1996) were first raised by Linebarger (1980, 1987), to argue that "weakly" negative expressions do not license NPIs directly but only in virtue of negative implicatures that they contribute to the communicative context.

(22) a. Exactly four people in the whole room *budged an inch* when I asked for help.

b. He kept writing novels long after he had *any* reason to believe that they would sell. (Ladusaw, 1996, p. 329)

Pragmatics was eventually agreed to be a crucial part of the story not only in the solution of that second problem, 'contextual licensing of NPIs', but also in the analysis of the content of the polarity items themselves. Kadmon and Landman (1990) argued for a unified treatment of NPI *any* and free choice *any* (as in *Any doctor will recommend more exercise*), and for the need to supplement Ladusaw's account with more about the meaning of *any* itself. They argued that the meaning of *any* is like that of *a* plus semantic/pragmatic conditions that reduce tolerance for exceptions: "widening", and "strengthening". Michael Israel (1996, 1998) has more recently extended that kind of account with further explicit pragmatics. I am omitting details, but I mention this to point out one important domain where the analysis of quantifiers (and in this case, not only quantifiers, although the determiners *every*, *any*, and *no* have played a leading role) has figured in a major turning point in the history of formal semantics, namely the recognition that formal semantics and formal pragmatics have to be studied together. One cannot just do semantics in isolation, and then take pragmatics to involve just the interaction of semantically interpreted sentences with the contexts in which they are used.

Quantifiers also starred in work that led to some major rethinking of the relation between semantics and pragmatics in the 1980's and the move to dynamic semantics. (These developments are described more fully in several places, including (Partee & Hendriks, 1997).) The work of Kamp and Heim beginning in the early 1980's was one of the major developments in the semantics of noun phrases, quantification, and anaphora. And more generally, their work influenced the shift from a "static" to a "dynamic" conception of meaning.

At the time of their work, indefinites like *a student* had been puzzling for a long time. Frege didn't treat them. Russell analyzed sentences containing indefinites as quantified sentences with an existential quantifier. Early work in formal semantics by Montague, Barwise and Cooper, and others absorbed that view of indefinites into Generalized Quantifier theory, analyzing *a student* as in (23).

(23) *A student* $\quad\quad \lambda P \exists x [\textbf{student}(x) \& P(x)] \quad\quad$ type $\langle\langle e, t\rangle, t\rangle$

But indefinites do not behave like other quantifier phrases.

(i) Singular indefinites permit discourse anaphora, unlike *every boy, no boy*.

(24) a. A boy came in. He was whistling.

b. #No/every boy came in. He was whistling.

(ii) Singular indefinites have 'variable quantificational force' and figure in puzzling anaphoric relations in Geach's famous donkey sentences (Geach, 1962).

(25) a. Every farmer who owns a donkey beats it.

b. If a farmer owns a donkey, he always beats it.

Kamp (1981) and Heim (1982, 1983) offered solutions to these classic problems; on their theories, indefinite noun phrases are interpreted as variables (in the relevant argument position) plus open sentences, rather than as quantifier phrases. The puzzle about why an indefinite NP seems to be interpreted as existential in simple sentences but universal in the antecedents of conditionals stops being localized on the noun phrase itself. Its apparently varying interpretations are explained in terms of the larger properties of the structures in which it occurs, which contribute explicit or implicit unselective binders that bind everything they find free within their scope.

Both Kamp and Heim make a major distinction between quantificational and non-quantificational NPs; the semantics of indefinites, definites, pronouns and names is on their analysis fundamentally different from the semantics of "genuinely quantificational" NPs like *every student*. The diversification of NP semantics, and the unification of some kinds of determiner quantification with some kinds of adverbial quantification in examples like (25a-b), represented an important challenge to Montague's uniform assignment of semantic types to syntactic categories, and in particular to the uniform treatment of NPs as generalized quantifiers. That challenge motivated work on type shifting and type-driven interpretation (Klein & Sag, 1985; Partee, 1986); and Kamp's challenges to compositionality in his Discourse Representation Theory led first to the competing development of "Dynamic Montague Grammar" (Groenendijk & Stokhof, 1990, 1991) and then to a resolution (Muskens, 1993).

## 6 Quantifiers, Universals, and Typology

There is much more to say about quantifiers in the history of formal semantics. Much of the early history that I have described is very directly concerned with the relation of logic to language, and to the interplay between new ideas in linguistic theory and new ideas and new formal tools in logic that together have helped to bridge the seeming disparities between "logical form" and "linguistic form". I will close with a few remarks about later developments more internal to linguistics, as formal semantics has become sufficiently "naturalized" within the field to be applied to such traditional linguistic concerns as language typology and the question of how much in language is universal and in what ways languages differ from one another.

The study of universals and typology is much farther advanced in phonology, morphology, and syntax than in semantics, but formal semantics has made advances in that domain in recent decades, and there have probably been more advances in the study of quantification than in any other area.

(Bach, Jelinek, Kratzer, & Partee, 1995) was the first major work on typology from the perspective of formal semantics. One of the questions that motivated our work was whether all natural languages have NPs that are interpreted as generalized quantifiers. Barwise and Cooper (1981) had hypothesized "Yes"; we marshaled our colleagues to help us answer the question, and it turned out to be "No". At least as widespread, but maybe also not universal, is "adverbial quantification" as in (26) and (25b), first studied by David Lewis (1975), and important in the work of Kamp and Heim.

(26) A quadratic equation usually has two distinct roots.

Some of the earliest work on semantic universals was Barwise and Cooper's work on Determiner universals, discussed above. Since that time, there has been much more work on universals and typology of determiners and quantifiers, including several recent important works (Keenan & Paperno, 2012; Peters & Westerståhl, 2006; Szabolcsi, 2010).

I have left many exciting recent developments unmentioned; as history moves closer to the present, research multiplies and one can hardly keep up with it, much less discuss it in a short article. It was not surprising that quantification was one of the first topics to be explored by linguists, logicians and philosophers working together. What may be more surprising is that even as formal semantics expands its reach into many more aspects of language, research on quantification is as active and innovative as ever.

## References

Abney, S. (1987). *The English noun phrase in its sentential aspect.* Unpublished doctoral dissertation, MIT, Cambridge, MA.
Ajdukiewicz, K. (1960). *Język i poznanie (Language and knowledge).* Warsaw.
Bach, E. (1968). Nouns and noun phrases. In E. Bach & R. Harms (Eds.), *Universals in linguistic theory* (pp. 90–122). New York: Holt, Rinehart.
Bach, E., Jelinek, E., Kratzer, A., & Partee, B. H. (Eds.). (1995). *Quantification in natural languages.* Dordrecht: Kluwer.
Bar-Hillel, Y. (1954). Logical syntax and semantics. *Language, 30,* 230–237.
Barwise, J., & Cooper, R. (1981). Generalized quantifiers and natural language. *Linguistics and Philosophy, 4,* 159–219.
Burris, S. (2009). The algebra of logic tradition. In E. N. Zalta (Ed.), *The Stanford encyclopedia of philosophy (summer 2009 edition).* Stanford: Stanford University.

Carnap, R. (1956). *Meaning and necessity: A study in semantics and modal logic. 2nd edition with supplements*. Chicago: Phoenix Books, University of Chicago Press.

Chomsky, N. (1955). Logical syntax and semantics: Their linguistic relevance. *Language, 31*, 36–45.

Chomsky, N. (1957). *Syntactic structures*. The Hague: Mouton.

Chomsky, N. (1965). *Aspects of the theory of syntax*. Cambridge, MA: MIT Press.

Church, A. (1951). A formulation of the logic of sense and denotation. In P. Henle, H. Kallen, & S. Langer (Eds.), *Structure, method and meaning: essays in honor of H. M. Sheffer*. New York: Liberal Arts Press.

Cocchiarella, N. (1997). Formally-oriented work in the philosophy of language. In J. Canfield (Ed.), *Philosophy of meaning, knowledge and value in the 20th century* (pp. 39–75). London; New York: Routledge.

Feferman, S. (2004). Tarski's conceptual analysis of semantical notions. In A. Benmakhlouf (Ed.), *Sémantique et épistémologie* (pp. 79–108). Paris: Casablanca Editions Le Fennec [distrib. J. Vrin].

Geach, P. (1962). *Reference and generality*. Ithaca: Cornell University Press.

Groenendijk, J., & Stokhof, M. (1990). Dynamic Montague grammar. In L. Kálman & L. Pólos (Eds.), *Papers from the second symposium on logic and language* (pp. 3–48). Budapest: Adakémiai Kiadó.

Groenendijk, J., & Stokhof, M. (1991). Dynamic predicate logic. *Linguistics and Philosophy, 14*, 39–100.

Harris, R. (1993). *The linguistics wars*. New York and Oxford: Oxford University Press.

Heim, I. (1982). *The semantics of definite and indefinite noun phrases*. Unpublished doctoral dissertation, University of Massachusetts, Amherst; published 1989, New York: Garland.

Heim, I. (1983). File change semantics and the familiarity theory of definiteness. In R. Bäuerle, C. Schwarze, & A. von Stechow (Eds.), *Meaning, use and the interpretation of language* (pp. 164–190). Berlin: Walter de Gruyter.

Hodges, W. (1985/6). Truth in a structure. *Proceedings of the Aristotelian Society, 86*, 135–151.

Israel, M. (1996). Polarity sensitivity as lexical semantics. *Linguistics and Philosophy, 19*(6), 619–666.

Israel, M. (1998). *The rhetoric of grammar: Scalar reasoning and polarity sensitivity*. Unpublished doctoral dissertation, University of California at San Diego.

Kadmon, N., & Landman, F. (1990). Polarity sensitive *any* and free choice *any*. In M. Stokhof & L. Torenvliet (Eds.), *Proceedings of the Seventh Amsterdam Colloquium* (pp. 227–252). ITLI: University of Amsterdam.

Kalish, D., & Montague, R. (1964). *Logic: Techniques of formal reasoning*. New York: Harcourt, Brace & World, Inc.

Kamp, H. (1981). A theory of truth and semantic representation. In J. Groenendijk, T. Janssen, & M. Stokhof (Eds.), *Formal methods in the study of language; Mathematical Centre tracts 135* (pp. 277–322). Amsterdam: Mathematical Centre. (Reprinted in: J. Groenendijk, Th. Janssen, and M. Stokhof (Eds.), 1984, Truth, Interpretation, Information, GRASS 2, Dordrecht: Foris, pp. 1–41)

Kaplan, D. (1964). *Foundations of Intensional Logic*. Unpublished doctoral dissertation, University of California Los Angeles: Ph.D. Dissertation.

Karttunen, L. (1968). *What do referential indices refer to?* Santa Monica, CA: The Rand Corporation.

Karttunen, L. (1969). Pronouns and variables. In *CLS 5* (pp. 108–116). Chicago: Chicago Linguistic Society, University of Chicago.

Katz, J. J., & Fodor, J. A. (1963). The structure of a semantic theory. *Language, 39*, 170–210.

Katz, J. J., & Postal, P. M. (1964). *An integrated theory of linguistic descriptions*. Cambridge, MA: MIT Press.

Keenan, E. L. (2003). The definiteness effect: semantic or pragmatic? *Natural Language Semantics, 11*(2), 187–216.

Keenan, E. L., & Paperno, D. (Eds.). (2012). *Handbook of quantifiers in natural language*. Springer.

Keenan, E. L., & Westerståhl, D. (1997). Generalized quantifiers in linguistics and logic. In J. van Benthem & A. ter Meulen (Eds.), *Handbook of logic and language* (pp. 837–893). Amsterdam: Elsevier Science B.V.

Klein, E., & Sag, I. A. (1985). Type-driven translation. *Linguistics and Philosophy, 8*, 163–201.

Ladusaw, W. A. (1979). *Polarity sensitivity as inherent scope relations*. Unpublished doctoral dissertation, University of Texas at Austin.

Ladusaw, W. A. (1996). Negation and polarity items. In S. Lappin (Ed.), *The handbook of contemporary semantic theory* (pp. 321–341). Oxford: Blackwell.

Lakoff, G. (1968). *Pronouns and reference. Parts I and II.* Bloomington: Indiana University Linguistics Club.

Larson, R. (1995). Semantics. In L. Gleitman & M. Liberman (Eds.), *An invitation to cognitive science. Vol 1: Language* (2nd ed., pp. 361–380). Cambridge, MA: The MIT Press.

Lewis, D. (1970). General semantics. *Synthése, 22*, 18–67.

Lewis, D. (1975). Adverbs of quantification. In E. L. Keenan (Ed.), *Formal semantics of natural language* (pp. 3–15). Cambridge: Cambridge University Press.

Lindström, P. (1966). First order predicate logic with generalized quantifiers. *Theoria, 32*, 186–195.

Linebarger, M. C. (1980). *The grammar of negative polarity.* Unpublished doctoral dissertation, MIT, Cambridge, MA.

Linebarger, M. C. (1987). Negative polarity and grammatical representation. *Linguistics and Philosophy, 10*, 325–387.

McCawley, J. (1971). Where do noun phrases come from? In D. Steinberg & L. Jakobovits (Eds.), *Semantics. An interdisciplinary reader in philosophy, linguistics and psychology* (pp. 217–231). Cambridge: Cambridge University Press.

Montague, R. (1968). Pragmatics. In R. Klibanski (Ed.), *Contemporary philosophy* (pp. 102–121). Florence: La Nuova Italia Editrice. (Reprinted in Montague (1974), 95–118)

Montague, R. (1970). Pragmatics and intensional logic. *Synthèse, 22*, 68–94. (Reprinted in Montague (1974), 119–147)

Montague, R. (1973). The proper treatment of quantification in ordinary English. In K. J. J. Hintikka, J. M. E. Moravcsik, & P. Suppes (Eds.), *Approaches to natural language* (pp. 221–242). Dordrecht: Reidel. (Reprinted in Montague (1974), 247–270)

Montague, R. (1974). *Formal philosophy. Selected papers of Richard Montague.* New Haven/London: Yale University Press. (Edited and with an introduction by Richmond H. Thomason)

Mostowski, A. (1957). On a generalization of quantifiers. *Fundamenta Mathematica, 44*, 12–36.

Muskens, R. (1993). A compositional discourse representation theory. In P. Dekker & M. Stokhof (Eds.), *Proceedings of the 9th Amsterdam Colloquium* (pp. 467–486). Amsterdam: ILLC, University of Amsterdam.

Newmeyer, F. J. (1980). *Linguistic theory in America: The first quarter-century of transformational generative grammar.* New York: Academic Press.
Newmeyer, F. J. (1996). *Generative linguistics : A historical perspective.* London; New York: Routledge.
Niiniluoto, I. (1994). Defending Tarski against his critics. In B. Twardowski & J. Wolenski (Eds.), *Sixty years of Tarski's definition of truth* (pp. 48–68). Warsaw: Philed.
Parsons, T. (1968). *A semantics for English.* (Unpublished paper.)
Parsons, T. (1972). *An outline of a semantics of English.* University of Massachusetts: Amherst.
Partee, B. H. (1973). Some transformational extensions of Montague grammar. *Journal of Philosophical Logic, 2,* 509–534.
Partee, B. H. (1975). Montague grammar and transformational grammar. *Linguistic Inquiry, 6,* 203–300.
Partee, B. H. (Ed.). (1976). *Montague grammar.* New York: Academic Press.
Partee, B. H. (1986). Noun phrase interpretation and type-shifting principles. In J. Groenendijk, D. de Jongh, & M. Stokhof (Eds.), *Studies in discourse representation theory and the theory of generalized quantifiers* (pp. 115–143). Dordrecht: Foris.
Partee, B. H. (2011). Formal semantics: Origins, issues, early impact. In B. H. Partee, M. Glanzberg, & J. Skilters (Eds.), *Formal semantics and pragmatics. Discourse, context, and models. The Baltic international yearbook of cognition, logic, and communication. Vol. 6 (2010)* (pp. 1–52). Manhattan, KS: New Prairie Press.
Partee, B. H., & Hendriks, H. L. W. (1997). Montague grammar. In J. van Benthem & A. ter Meulen (Eds.), *Handbook of logic and language* (pp. 5–91). Amsterdam/Cambridge, MA: Elsevier/MIT Press.
Peters, S., & Westerståhl, D. (2006). *Quantifiers in language and logic.* Oxford: Oxford University Press.
Rosenbaum, P. S. (1967). *The grammar of English predicate complement constructions.* Cambridge, MA: MIT Press.
Russell, B. (1957). Mr. Strawson on referring. *Mind, 66*(263), 385–389.
Scott, D. (1970). Advice on modal logic. In K. Lambert (Ed.), *Philosophical problems in logic. some recent developments* (pp. 143–174). Dordrecht: Reidel.
Stanley, J. (2008). Philosophy of language in the twentieth century. In D. Moran (Ed.), *The Routledge companion to twentieth century phi-*

*losophy* (pp. 382–437). London: Routledge Press.

Stokhof, M. (2006). The development of Montague grammar. In S. Auroux, E. F. K. Koerner, H.-J. Niederehe, & K. Versteegh (Eds.), *History of the language sciences* (Vol. 3, pp. 2058–2073). Berlin—New York: Walter de Gruyter.

Strawson, P. F. (1950). On referring. *Mind, 59*, 320–344.

Szabolcsi, A. (2010). *Quantification*. Cambridge: Cambridge University Press.

Tarski, A. (1935). Der Wahrheitsbegriff in den formalisierten Sprachen. *Studia Philosophica, 1*, 261–405.

Westerståhl, D. (2011). Generalized quantifiers. In E. N. Zalta (Ed.), *The Stanford encyclopedia of philosophy (summer 2011 edition)*.

Barbara H. Partee
Department of Linguistics
University of Massachusetts
Amherst, MA 01003, USA
e-mail: partee@linguist.umass.edu
URL: http://people.umass.edu/partee/

# Wittgenstein Completeness

## VICTOR RODYCH

**Abstract:** The contention that there are *unanswerable questions* arose, in the 19th Century, after much scientific and mathematical success. In 1880, Emil du Bois-Reymond declared that "we do not know and we will not know"—that there are indeed questions that cannot be answered. 20 years later, David Hilbert responded that "in mathematics there is no ignorabimus"—that any mathematical problem can be solved. In his *Tractatus*, Ludwig Wittgenstein similarly rejected du Bois-Reymond's pessimism by stating that "*[t]he riddle* does not exist" and explaining why. This paper examines the issues of solvability, decidability, completeness, and incompleteness by looking at the context of Hilbert's initial assertion, the views of Hilbert, Brouwer, Wittgenstein and Skolem in the crucial years 1889-1930, and Wittgenstein's rejection of mathematical undecidability both before and after Gödel's 1931 undecidability result. The sketch here given aims to show (1) how Wittgenstein was innervated by Brouwer's 1928 lecture to reject Brouwer's undecidable propositions, (2) why Brouwer expects incompleteness, why Hilbert expects completeness, and why Wittgenstein rejects the *possibility* of incompleteness, despite an agreement about the inexhaustibility of mathematics, and (3) that Wittgenstein's internally coherent view is externally plausible in part because *incompleteness* has not been proved.

**Keywords:** decidability, incompleteness, Wittgenstein, Hilbert, Brouwer, Gödel

Are there questions that are *unanswerable*?

The growth and enormous success of empirical science seems to have led to a general discussion in the 19th Century about possible *limits* of scientific inquiry and related questions as to the proper jurisdictions for questions of various types.[1] This culminated in Emil du Bois-Reymond's 1880 address to the Berlin Academy of Sciences, where du Bois-Reymond answered the afore-mentioned question in the affirmative, stating that some questions or 'riddles' *cannot be answered*. Du Bois-Reymond listed three such questions

---

[1] It is worth noting that near the end of the 20th Century, in his *Rocks of Ages* (1999), Stephen Jay Gould argued for "non-overlapping magisteria" (NOMA) of "matters of fact" (p. 53) vs. "moral issues about the value and meaning of life," which has since been criticized by especially Richard Dawkins, who, in his *The God Delusion*, argues that the existence of God is a *scientific question* which has *received* a strongly negative answer (Dawkins, 2006, pp. 46–50).

(i.e., the ultimate nature of matter or force, the origin of motion, the origin of simple sensory sensation and consciousness,) and concluded that "we do not know and we will not know" ("ignoramus et ignorabimus").

In 1899, in his *Die Welträthsel* (*World Riddle*), Ernst Haeckel explicitly stated that du Bois-Reymond's 'idea of the "seven great enigmas" has been widely accepted.' Haeckel, however, *rejected* du Bois-Reymond's pessimism, asserting that the three allegedly "transcendental and insoluble" problems "are settled by our conception of substance," that the three "difficult riddles" "are decisively answered by our modern theory of evolution," and that "the freedom of the will" is a "pure dogma, based on an illusion, and has no real existence."[2]

One year later, at the International Congress of Mathematicians at Paris in 1900, David Hilbert replied to the skepticism of du Bois-Reymond by delineating fundamental mathematical problems and famously declaring that "in mathematics there is no *ignorabimus*."

About 20 years after Hilbert's declaration, near the end of his *Tractatus Logico-Philosophicus*, Ludwig Wittgenstein declares:[3]

> When the answer cannot be put into words, neither can the question be put into words. *The riddle* does not exist. If a question can be framed at all, it is also *possible* to answer it. (6.5)
>
> Scepticism is not irrefutable, but obviously nonsensical, when it tries to raise doubts where no questions can be asked. For doubt can exist only where a question exists, a question only where an answer exists, and an answer only where something *can be said*. (6.51)[4]

It is interesting to see both Hilbert and Wittgenstein replying to du Bois-Reymond's *skeptical* conception of *unanswerable riddles*, a concep-

---

[2] Cf. (Haeckel, 1899, p. 16). Haeckel speaks of the "unnatural and fatal opposition between Science and Philosophy" (Preface, XII), adding that Philosophy "is far from assimilating the hard-earned treasures of empirical science" and that 'representatives' of "exact science" are equally guilty of ignoring the 'deeper' concerns of Philosophy. The result is that there is still not yet a "Philosophy of Nature," which Haeckel offers in his Monistic Philosophy.

[3] See (Wittgenstein, 1921).

[4] Though not germane to the focus of this paper, it should be mentioned that Wittgenstein *does* say that "[t]here are, indeed, things that cannot be put into words"—"[t]hey *make themselves manifest*"—"[t]hey are what is mystical." Then he famously ends the *Tractatus* by asserting that there are *no philosophical propositions*, only "propositions of natural science" (6.53), namely, contingent propositions.

tion which, according to Haeckel, had "been widely accepted" up to the time of Hilbert's 1900 response in Paris.

When Hilbert first declared "there is no *ignorabimus*" in mathematics in 1900, he meant that where there is a mathematical problem or question, the mathematician can, *in principle*, find a proof that will solve the problem or answer the question—there are *no absolutely undecidable* mathematical problems.[5] As we shall see, in 1900 Hilbert *also* inaugurates an *axiomatic conception of mathematical completeness* saying that an axiom system for the real numbers must constitute "a complete description of the relations subsisting between" real numbers and, most importantly, that if such a system of axioms is consistent, all and only *derivable propositions* are true (i.e., the system is both sound and complete) (Hilbert, 1902, pp. 447–448).

In this paper I will sketch Wittgenstein's reasons for rejecting this conception of mathematics and, with it, questions concerning the completeness or incompleteness of *any* axiomatic mathematical calculus. As we shall see, Wittgenstein rejects the very possibility of an incomplete mathematical calculus *before* and *after* Gödel's 1931 paper. On Wittgenstein's account, all mathematical calculi are *complete* because we can (algorithmically) decide all genuine mathematical propositions, namely expressions *for which we knowingly have an applicable and effective decision procedure*. In this respect, Wittgenstein's post-1928 answer to Hilbert's demand for completeness is to simultaneously reject *unanswerable* "mathematical riddles" *and* Hilbert's demand for a completeness proof. The nature of mathematics ensures mathematical completeness, according to Wittgenstein, by denying the distinction between syntax and semantics required by any alleged incompleteness (result).

## 1  Hilbert on Solvability, Decidability, and Completeness in 1900

20 years after du Bois-Reymond concluded that "we do not know and we will not know" ("ignoramus et ignorabimus"), Hilbert replied to du Bois-Reymond's skepticism by asserting:

> the conviction (which every mathematician shares, but which no one has as yet supported by a proof) that every definite mathematical problem must necessarily be susceptible of an exact

---
[5]The question whether any mathematical problem is *absolutely undecidable* is first discussed in 1907 by Brouwer and then later by Skolem and Gödel. See section 4, below.

> settlement, either in the form of an actual answer to the question asked, or by the proof of the impossibility of its solution and therewith the necessary failure of all attempts. ...
>
> This conviction of the solvability of every mathematical problem is a powerful incentive to the worker. We hear within us the perpetual call: There is the problem. Seek its solution. You can find it by pure reason, for in mathematics there is no *ignorabimus*. (Hilbert, 1902, p. 445)

Hilbert would make this famous declaration at least three more times,[6] lastly at his retirement address in Königsberg, September 8th, 1930, where he ends by saying that "for the mathematician there is no ignorabimus, nor, in my opinion, for any part of natural science," and that, therefore, "our answer is... we must know, we shall know" (Hilbert, 1930, p. 1165)![7]

It is worth noting that in 1900, immediately after saying that "in mathematics there is no *ignorabimus*," Hilbert adds that "[t]he supply of problems in mathematics is inexhaustible"! Thus, at the very beginning in 1900, there is a *tension* between, on the one hand, the *solvability* of any mathematical problem and the *completeness* of a given mathematical system, and, on the other hand, the *inexhaustibility* of mathematical problems. Hilbert's first point is that *any given* mathematical problem can, *in principle*, be solved. There are, therefore, no *absolutely unsolvable mathematical problems*. Hilbert's second point, however, is unclear; he may mean that there are non-denumerably many mathematical problems (and we can only ever solve finitely many).

There is no question that Hilbert has at this time an *axiomatic conception of mathematical completeness*. In stating problem #2 as the problem of providing an absolute consistency proof for arithmetic, Hilbert says that (1) "in investigating the foundations of a science, we *must* set up a system of axioms which contains an exact and *complete* description of the relations subsisting between the elementary ideas of that science" (emphases mine), (2) "[t]he axioms so set up are at the same time the definitions of those elementary ideas," and (3) "no statement within the realm of the science whose foundation we are testing is held to be correct unless it can be derived from those axioms by means of a finite number of logical steps."

In the case before us, where we are concerned with the axioms

---
[6] See also (Hilbert, 1925, p. 384) and (Hilbert, 1929, p. 233).
[7] These concluding words are engraved on Hilbert's tombstone.

of real numbers in arithmetic, the proof of the compatibility of the axioms is at the same time the proof of the mathematical existence of the complete system of real numbers or of the continuum. The totality of real numbers, . . ., is not the totality of all possible series in decimal fractions, [but] rather a system of things whose mutual relations are governed by the axioms set up and for *which all propositions, and only those, are true which can be derived from the axioms by a finite number of logical processes*. [Emphasis mine.] (Hilbert, 1902, pp. 447–448)

Though this statement is not as sophisticated, informed, or as concrete as Hilbert's late 1920's statements,[8] Hilbert clearly says here, in 1900, that an axiom system for the real numbers must constitute "a complete description of the relations subsisting between" real numbers and, most importantly, that if such a system of axioms is consistent, all and only *derivable propositions* are true (i.e., the system is both sound and complete).

Thus, one year after axiomatizing Euclidean Geometry in his *The Foundations of Geometry*, Hilbert states "that every definite mathematical problem must necessarily be susceptible of an exact settlement" and by this he *seemingly means* that, in principle, we must be able to solve any and all mathematical problems by axiomatizing each and every branch of mathematics. The afore-mentioned tension is lurking here, however, for if Hilbert means that we must be able to axiomatize *all* of mathematics by sound and complete axiomatic calculi, this would contradict Hilbert's explicit claim that "the supply of problems in mathematics is inexhaustible." However this important tension might be resolved,[9] Hilbert *minimally* means that we can axiomatize elementary number theory and the theory of real numbers such that those axiomatic systems are sound and complete.

---

[8] See (Hilbert, 1929, pp. 229–233) for Hilbert's famous 1928 statement of his four problems of Proof Theory. Wang says (Wang, 1987, p. 81) that Hilbert articulates the following four problems: "Roughly speaking they are (1) finitist consistency proof of analysis, (2) that of set theory, (3) completeness of first-order number theory and analysis, (4) completeness of first-order logic." Hilbert declares (Hilbert, 1929, p. 233) that the 'scepticism' then "not infrequently expressed against science" could only be countered if *in mathematics* "there was certain truth." "Proof theory makes such an attitude impossible and brings us the exaltation of the conviction that at least the mathematical understanding encounters no limits and that it is even capable of discovering the laws of its own thought."

[9] Perhaps the most plausible way for Hilbert to dissolve this tension is to argue that mathematics is always expanding and that it is indefinitely extensible. Such an argument, however, might have to restrict intuition to certain core mathematical systems—such as elementary number theory—which would raise the question of what *demarcates* a mathematical system from a merely formal, non-mathematical system.

*Why* does Hilbert adopt this position? It is, I believe, due to his life-long *Kantian* belief in the importance of fundamental, *a priori* intuitions for both geometry and arithmetic (a view shared by Bertrand Russell in the crucial period 1900-1913 (Floyd & Dreben, 1991, p. 28)). As early as *The Foundations of Geometry*, Hilbert says that "[t]he axioms of geometry can be divided into five groups" and that "[e]ach of these groups expresses certain related facts basic to our *intuition*" [emphasis mine] (Hilbert, 1899, p. 3). In "On the Infinite," Hilbert still maintains the crucial importance of Kantian *a priori* intuition in mathematics, claiming that Frege's and Dedekind's attempts "to make pure logic provide for arithmetic a foundation that would be independent of all intuition and experience" "were bound to fail" because as "Kant [correctly] taught... mathematics has at its disposal a content secured independently of all logic and hence can never be provided with a foundation by means of logic alone" (Hilbert, 1925, p. 376; Hilbert, 1930, p. 1161). Hilbert's belief in *a priori* intuition compels him to (a) think of a mathematical *domain*, such as the natural numbers, as a *determinate* structure *about which we can make clear and 'definite'* claims (i.e., number-theoretic propositions), (b) believe also that *intuition* enables us to lay down syntactical rules for the well-formedness of propositions about the *fully determinate structure* of the natural numbers, and (c) axiomatize a particular mathematical domain or intuitive system. Put differently, there are determinate mathematical facts which can be discovered by means of "pure reason" and false propositions whose falsity can be proved by means of *impossibility proofs*.

Hilbert could not have even claimed that every mathematical problem is solvable or that axiomatic mathematical calculi must be complete (and sound) without having a distinction between mathematical *language*, mathematical *meaning*, and mathematical *reality* (or *domains*, or *structures*). On his view, before an axiomatic system is constructed, we *know by intuition* what constitutes a mathematical problem (and proposition) and what would constitute its solution (or decision). It is intuition that guides us in stipulating the appropriate syntax for a particular mathematical domain and it is also intuition that guides us in recognizing completeness and determining whether questions of axiomatic completeness have been rigorously decided.

## 2 Propositions, Truth, and Mathematics in the *Tractatus*

Wittgenstein's post-1928 views on mathematics, mathematical completeness and mathematical decidability are really elaborations of his 1918-21 views on propositions, truth and mathematics as expressed in the *Tractatus*.

In the *Tractatus*, Wittgenstein asserts that *only contingent propositions* have sense and, therefore, only they are true or false. Tautologies and contradictions are *senseless* and are only 'true' and 'false,' respectively, in very different senses from contingent propositions.[10] Mathematical pseudo-propositions (6.2) are *equations*, which are either 'correct' or 'incorrect' (i.e., *not* true or false). What tautologies, contradictions and mathematical equations have in common is that, *unlike contingent propositions*, which are true or false because of agreement or lack of agreement with reality, logical propositions and mathematical equations are determined to be correct or incorrect *syntactically*, "from the symbol alone" (6.113). From this point on, Wittgenstein resolutely eliminates truth and falsity from mathematics, speaking instead of proofs, decisions, decision procedures, calculations, and syntax.

## 3 Completeness In and After the *Tractatus*

A central part of the *Tractatus* is concerned with *completeness*. Wittgenstein explicitly works out a theory of an elementary (atomic) proposition that enables him to say what would constitute a *complete description*—a *complete picture*—of "the world."

> If all true elementary propositions are given, the result is a complete description of the world. The world is completely described giving all elementary propositions, and adding which of them are true and which false. (4.26)
>
> The totality of true thoughts is a picture of the world. (3.01)
>
> The totality of true propositions is the whole of natural science (or the whole corpus of the natural sciences). (4.11)[11]

---

[10]"Tautology and contradiction are the limiting cases—indeed the disintegration—of the combination of signs" (4.466).

[11] At (4.52), Wittgenstein adds that "[p]ropositions comprise all that follows from the totality of all elementary propositions." Given that propositions are either elementary propositions, or truth-functions of elementary propositions, this, too, rules out logical or mathematical 'propositions' as genuine propositions.

We can give a *complete* description of the world by giving all true elementary propositions *because* "[t]he totality of existing states of affairs is the world" (2.04) and each true elementary proposition agrees with exactly one existent state of affairs.

From this conception it follows that *there are no (absolutely) unanswerable or undecidable questions about the world*—no questions about the world that are *in principle* unanswerable—because the world, though constantly changing, always consists of the totality of existing states of affairs. To ask a question about the world is, for Wittgenstein, to ask of a particular proposition, atomic or compound, whether it is true or false. In one of his most famous philosophical passages, ascending to the climax of the *Tractatus* and its rejection of *philosophical propositions*, Wittgenstein rejects skepticism and *unanswerable riddles*. He says, as we have already seen, that "[*t*]*he riddle* does not exist" and that "[i]f a question can be framed at all, it is also *possible* to answer it" (6.5). Conversely, if something can actually be *said*—if we can make a *claim* that such-and-such *is the case* by asserting a proposition—that proposition must either agree or disagree with reality. It is clear, therefore, that all meaningful questions about the world must *have answers in principle*.

As for *incompleteness* in the *Tractatus*, something is only a description if it is complete. There is no such thing as incomplete meaning or, more precisely, "incomplete sense." For Wittgenstein, "[t]he requirement that simple signs be possible is the requirement that sense be determinate" (3.23), and this means, in part, that every elementary proposition must have a *complete sense* (or a *completely determinate* sense). There can be no vagueness or incompleteness in an elementary proposition. The only way in which an elementary proposition can be *determinately true or false*—independent of our knowledge or ignorance—is if its sense is *fully determinate*. As Wittgenstein says, "[a] proposition is not a blend of words," it "is articulate" (3.141).

Determinacy of sense for elementary and compound propositions is connected with Wittgenstein's conception of a *complete description* of the world. If, e.g., elementary propositions could be indeterminate in the sense of not having complete and precise senses, then the totality of all constructible propositions *would not necessarily* enable us to completely describe the world. There might be states of affairs or facts *that we could not represent*. The only way to ensure that the world is completely describable—and therefore completely knowable in principle—is to ensure that an elementary proposition is completely determinate. Once that is ensured, Wittgenstein can ensure that all possible elementary propositions are constructible given the internal properties of objects and their respective *possible* states of affairs. This ensures that it is possible to give a complete description of the world for any particular time-slice.

## 4 Decidability and Completeness Between 1900 and 1928

Though the first modern completeness proofs—for the Propositional Calculus—were not executed until 1918 (Paul Bernays) and 1921 (Emil Post), these proofs and the propositional calculus itself provided a clear example of an axiomatic calculus in which semantics and syntax can be clearly separated.[12] Similarly, on Hilbert's emerging view between 1900 and 1928, the question of mathematical completeness is relative to an axiomatic calculus for which we *intuitively* have a clear distinction between its semantics and its syntax.[13] It is only because we believe we can stipulate rules for well-formedness such that the well-formedness of any string of symbols is decidable viz. a particular axiomatic calculus, that we *believe* we can pose and possibly answer the question of completeness for a particular axiomatic calculus.

L.E.J. Brouwer took issue with Hilbert's 1900 assertions as early as his 1907 doctoral dissertation.

> A fortiori it is not certain that any mathematical problem can either be solved or proved to be unsolvable, though HILBERT, in

---

[12]The question was probably immediate: Is First-Order Logic similarly complete? The initial answer was 'probably' or "we hope so." Indeed, in a letter to Bertrand Russell, dated November, 1913, Wittgenstein sketches a decision procedure for propositional logic and then asserts that "it is obvious" that a suitable decision procedure (i.e., *ab*-notation) "can be made up" to decide propositions of Russell's Theory of Apparent Variables (i.e., First-Order Logic). Interestingly, in a brief Dec. 28, 1930 discussion of Bernays' completeness proof with Schlick and Waismann, Wittgenstein seemingly rejects the possibility of proving the completeness of the Propositional Calculus or even comparing a complete Propositional Calculus and an *incomplete* "Propositional Calculus." Wittgenstein's then strong rejection of Hilbert's metamathematics was certainly operational—especially Wittgenstein's rejection of the need for, or possibility of, consistency proofs.

[13]It seems that Skolem was the first to conclude that the Continuum Problem may not be decidable in ZF (or ZFC). See (Skolem, 1923, p. 149, note 2): "Since Zermelo's axioms do not determine the domain B [the model for them], ... it is quite probable that what is called the continuum problem is not solvable at all on this basis". See also (Skolem, 1929, p. 222) and Gödel's 1947 conjecture (Gödel, 1947, p. 519) that the Continuum Hypothesis (CH) is "most likely" undecidable from the set-theoretical axioms, and that CH nonetheless "must be either true or false, and its undecidability from the axioms as known today can only mean that these axioms do not contain a complete description of this reality" (Gödel, 1947, p. 520).

> 'Mathematische Probleme', believes that every mathematician is deeply convinced of it. But for this question as well it is of course uncertain whether it will ever be possible to settle it, i.e. either to solve it or to prove that it is unsolvable (a logical question is nothing else than a mathematical problem). (Brouwer, 1907, p. 79)

One year later, Brouwer states, first, that "the question of the validity of the principium tertii exclusi is equivalent to the question *whether unsolvable mathematical problems exist*," second, that "[t]here is not a shred of a proof for the conviction... that there exist no unsolvable mathematical problems," and, third, that there are meaningful propositions/'questions,' such as "*Do there occur in the decimal expansion of π infinitely many pairs of consecutive equal digits?*", to which the Law of the Excluded Middle does not apply *because* "it must be considered as uncertain whether problems like [this] are solvable" (Brouwer, 1908, pp. 109–110).

Interestingly, around the time of Hilbert's articulation of the four main problems for Proof Theory, Thoralf Skolem similarly questioned the decidability of all number-theoretic problems.

> A very probable consequence of this relativism is again that it cannot be possible to completely characterize the mathematical concepts; this already holds for the concept of the natural number. Thereby arises the question, whether the unicity or categoricity of mathematics might not be an illusion. Then it would not at all be strange if some problems were unsolvable; they would in fact not be decided by means of the principles which we are able to found them with, and it would not at all be necessary to resort to a new logic, as Brouwer does, in order to see this. (Skolem, 1929, 1970, p. 224)

## 5 Wittgenstein on Mathematics: Syntax *without* Semantics

On March 10, 1928, though they did not meet, Gödel and Wittgenstein both attended L.E.J. Brouwer's Vienna lecture "Mathematics, Science, and Language" (Brouwer, 1929). "[A]ccording to Carnap," Hao Wang tells us, "[Gödel] was influenced by Brouwer's lecture to assert that mathematics is inexhaustible (by any formal system)" (Wang, 1987, pp. 80, 88). As for

Wittgenstein, there is certainly no doubt that he was greatly *energized* by Brouwer's lecture (Feigl, 1981; Menger, 1994), and it also seems probable that he was influenced by Brouwer's lecture and private communication (Finch, 1977, p. 260; van Dalen, 2005, pp. 566–567).[14] It seems probable that Wittgenstein's *thinking* was influenced (in a related but different direction) by Brouwer's explicit claims, *at this very lecture*, that mathematicians have mistakenly "derived logical principles [e.g., *"Excluded Middle"*] from the language describing finite mathematics and *without scruple* also used them in their pure mathematical study of *infinite systems*" and mathematicians were "led... to accept and trust assertions derived by means of the logical principles even when these *could not be subjected to direct check*" (Brouwer, 1929, pp. 49–50; emphases mine). Brouwer's emphasis on the *difference* between finite and infinite systems and his discussion of putatively undecidable propositions may well have prompted Wittgenstein, less than a year later, (1) to begin working out his crucial distinction between finite sets as *extensions* versus "infinite sets" as unlimited *rules* for generating ever-greater finite extensions, and to (2) to abandon his Tractarian conception of quantified propositions as (possibly *infinite*) conjunctions and disjunctions (Moore, 1955, pp. 2–3) and reject *infinitistic quantification* inside and outside of mathematics. For Wittgenstein, because there are no infinite mathematical extensions—existent or constructible to completion—he is driven to elaborate his Tractarian conception of rules by *identifying* infinite/unlimited rules with *mathematical infinity*.

What Wittgenstein had *before* hearing Brouwer's lecture was a *non-referential* conception of mathematics. From the *Tractatus* until the end of his life, Wittgenstein asserts that mathematical terms and 'propositions' do not refer to or talk about objects or things or a realm of existent entities. For Wittgenstein, contra Brouwer, mathematics is essentially *syntactical*, for all we actually have in mathematics are (finite) extensions and rules (intensions). Where Brouwer claims that symbols are *inessential* in mathematics because mathematics is essentially mental (Brouwer, 1929, p. 50), Wittgenstein claims that mathematics consists only of symbols and rules for operating on and with these symbols (and certain human behaviour and agreement). As Wittgenstein says: "In mathematics *everything* is algorithm and *nothing* is meaning ['Bedeutung']" (PG 468). Since there are no mathematical objects about which mathematical propositions speak, mathematical proofs can only establish *syntactical connections* among *signs* in

---

[14] See (Marion, 2003, p. 107), who may disagree about the influence or see it differently.

a mathematical calculus (e.g., among: $27, 9, 3, \times, =$). There is no place in mathematics for truth or falsity, and for this reason, Wittgenstein resolutely speaks of *formal, syntactical algorithms and derivations*. This does not mean that mathematicians are precluded from pursuing and trying to decide, e.g., Goldbach's Conjecture (GC). If, however, we prove GC by means of an *inductive* proof, we will have created a new calculus by extending our calculus with a newly proved inductive base and inductive step, from which we can prove "infinitely many" particular propositions of the form "GC(12,444)."

Brouwer and Hilbert both claim (a) that intuition is essential to mathematics and to mathematical decidability and (b) that mathematical terms and propositions have meaning and/or reference. Wittgenstein rejects both (a) and (b) and, therefore, rejects the standard conception of mathematical truth and falsity. On Wittgenstein's view, all that we can and should say is that some propositions *are derived* within, e.g., PA, and are therefore *derivable*, and others are not presently derivable or refutable within PA. Thus, where Brouwer sees *undecidable mathematical propositions* and the failure of the Law of the Excluded Middle, Wittgenstein sees *non*-mathematical expressions in mathematical guise and *undecidability itself* as an *indication* of something non-mathematical.

## 6 Wittgenstein on Decidability in Mathematics, 1929 and After

In 1929-30, in opposition to both Brouwer and Hilbert, Ludwig Wittgenstein works out a Philosophy of Mathematics whereby any formal mathematical calculus (system) is *necessarily complete*. A concatenation of signs $\alpha$ is a proposition of a calculus $\Gamma$ *iff* either (a) $\alpha$ has been proved or refuted in $\Gamma$ or (b) we know how to decide $\alpha$ in $\Gamma$ by means of an applicable and effective decision procedure. Thus, by stipulation, all mathematical calculi are *syntactically complete*. Let us call this "Wittgenstein Completeness."

> Meaningfulness = $\alpha$ is a proposition of calculus $\Gamma$ = we *know how* to algorithmically decide $\alpha$ in $\Gamma$.

We are mistaken, Wittgenstein argues, to think that propositions that quantify over, e.g., the infinite domain of natural numbers are meaningful (i.e., genuine mathematical propositions) because they are not algorithmically decidable. Consider, e.g., the following uncontroversial wffs.

"There exist infinitely many primes" (EPNT).

Goldbach's Conjecture (GC).

Fermat's Last Theorem (FLT).

"There do not exist five consecutive 7's in the decimal expansion of $\pi$." (PIC)

$\sim (\exists n) Bew(n, m)$ (Gödel's Undecidable Proposition).

Most mathematicians and philosophers claim that these putative propositions are *determinately true or false* in the absence of a decision *and* in the absence of a decision procedure. On Wittgenstein's view, however, this commits us to Platonism because, even on the received view, we don't even know whether "the wffs of PA" are decidable *in principle* in PA.

Interestingly, Saul Kripke reminds us that in the 1920's, Hilbert and Ackermann wanted to *eliminate* quantifiers using epsilon terms because, at that time, "quantification even over all the natural numbers" was "a dubious idea"![15] For Hilbert at the time: "it was thought that the subject of model theory... and the question of the completeness of quantification theory... was officially meaningless." In this connection, it is also worth noting that Alfred Tarski was his entire life a nominalist, who once said: "People have asked me, 'How can you, a nominalist, do work in set theory and logic, which are theories about things you do not believe in?'... I believe that there is value even in fairy tales and the study of fairy tales" (Feferman & Feferman, 2004, p. 52).

Even more surprisingly, especially given "the official story" by Gödel and others, is that *Gödel himself* espoused a similar *anti*-Platonism in 1933: "[O]ur axioms, if interpreted as meaningful statements, necessarily presuppose a kind of Platonism, which cannot satisfy any critical mind and which does not even produce the conviction that they are consistent" (Gödel, 1995, p. 19). And when, in the 1950's, Gödel made every effort to use his Incompleteness Theorems to refute "the syntactical viewpoint" of Carnap and Wittgenstein (*1953/9), he was *unsatisfied* with his arguments and refused to have the paper published in *The Philosophy of Rudolf Carnap* (1963).

The passage of time and the dominance of model theory and set theory have caused 2-3 generations to forget that and why Hilbert became a finitist

---
[15] S. Kripke, "The Collapse of the Hilbert Program," http://broadcast.iu.edu/ceremon/celeb07/index.html, Indiana University, October 15, 2007.

and just how radical his views became in the 1920s. The fact that Gödel was for a time an anti-Platonist and that Tarski was an anti-Platonist his entire life should give us pause: perhaps Wittgenstein's radical and unorthodox views are less radical than they seem. Certainly, in 1929-30 and even in 1937 Wittgenstein's views on unrestricted quantification were far less unorthodox than they are today.

## 7  Wittgenstein on Undecidability, 1929 and After

The combination of Wittgenstein's long-standing mathematical non-referentialism and his newer finitism and algorithmic decidability engender a new resolute rejection of mathematical undecidability.

> If someone says (as Brouwer does) that for $(x)f_1 x = f_2 x$, there is, as well as yes and no, also the case of undecidability, this implies that '$(x)\ldots$' is meant extensionally and we may talk of the case in which all $x$ happen to have a property. In truth, however, it's impossible to talk of such a case at all and the '$(x)\ldots$' in arithmetic cannot be taken extensionally. ...
>
> Undecidability presupposes that there is, so to speak, a subterranean connection between the two sides; that the bridge *cannot* be made with symbols. [E.g.: Between "$\sim (\exists n)\ Bew(n,m)$" and "$1, 2, 3, 4, \ldots$"]
>
> A connection between symbols which exists but cannot be represented by symbolic transformations is a thought that cannot be thought. If the connection is there, then it must be possible to see it. ...
>
> Of course, if mathematics were the natural science of infinite extensions of which we can never have exhaustive knowledge, then a question that was in principle undecidable would certainly be conceivable." (Wittgenstein, 1975, §174, 1929).

Wittgenstein's rejection of mathematical undecidability occurred *before* Gödel completed "On Formally Undecidable Propositions" in November of 1930. For example, in 1929-34 Wittgenstein repeatedly argues that pseudo-propositions such as (Brouwer's) "There do not exist three consecutive 7's in the decimal expansion of $\pi$" (PIC), are meaningless pseudo-propositions because they are not *algorithmically* decidable. Where Brouwer and

Gödel *saw reasons* to believe in the existence of undecidable mathematical propositions, Wittgenstein saw reasons to regard the undecidable as *non-mathematical*. Interestingly and ironically, Wittgenstein, Hilbert, Brouwer and Gödel all saw reasons to think that mathematics is *inexhaustible*.

Later, in 1937, after probably reading only the informal introduction to Gödel's (1931), Wittgenstein rejects the notion of a "true but unprovable proposition of *Principia Mathematica*" as a *contradiction-in-terms* and he discusses and rejects a semantic version of Gödel First Incompleteness Theorem similar to Gödel's own informal presentation in the introduction of his 1931 paper (Rodych, 1999, 2002, 2003, 2006). Wittgenstein presents what he takes to be a Gödelian Double *Reductio*. Someone says to Wittgenstein: "I have constructed a proposition (I will use '$P$' to designate it) in Russell's symbolism, and by means of certain definitions and transformations it can be so interpreted that it says: '$P$ is not provable in Russell's system'" (Wittgenstein, 1978, App. III, §8).

($P_1$) Either $P$ is true or $P$ is false.

($P_2$) Either $P$ is provable in PM or $P$ is not provable in PM.

($P_3$) If $P$ is false, then it is provable in PM, and if $P$ is provable in PM it is *true*, since all provable propositions in PM are true. (CONTRADICTION: False & True.)

($P_4$) If $P$ is provable in PM, then it is provable that $P$ is not provable in PM, since $P$ *means* or says "$P$ is not provable in PM." (CONTRADICTION: provable and unprovable.)

($C$) Thus, $P$ is true and not provable in PM.

Wittgenstein rejects this version of a proof—and the claim that Gödel has *proved incompleteness*—on the grounds that, for Wittgenstein *True in $\Gamma$ = Proved in $\Gamma$*. There cannot be a true proposition of PM that is not proved (or not provable) in PM. On Wittgenstein's view, since $P$ is not algorithmically decidable in PM, $P$ is simply not a proposition *of* PM. Moreover, P is *unusable* inside PM (or PA) and it is unusable outside PM (or PA) (e.g., in an application in Physics).

To those, like Mark Steiner (2001), who argue that Gödel's proof *demands* a forced extension of PM (or PA), Wittgenstein tacitly replies that we have *no reason* to add $P$ to PM on the grounds that $P$'s meaning *agrees with* what is true *and provable* in PM (i.e., $\sim F(1), \sim F(2), \sim F(3)$, etc.). These 'grounds' simply beg the question, since we cannot prove all of them.

Though the constraints of this paper do not permit an argument in support of Wittgenstein's highly unorthodox view of Gödel's syntactical proof, we should remind ourselves that what Gödel (and Rosser) actually *proved* is that if the PA fragment of PM is consistent, then Gödel's constructed formula $\varphi$ is *syntactically independent* of PA. That is to say, they proved:

> Either (1) *both* $\varphi$ and $\sim \varphi$ are derivable in PA OR (2) neither $\varphi$ nor $\sim \varphi$ is derivable in PA.

No one has proved that $\varphi$ is undecidable in PA. Moreover, no one has proved that $\varphi$ is *true* because it is not possible to prove *the provability of* $\sim F(1), \sim F(2), \sim F(3)$, etc., if PA is consistent. And since "all consistency proofs are relative consistency proofs" (Wang, 1987, p. 281), it is not possible to prove the incompleteness of PA in the intended sense.

On the received view, however, Gödel's result is an *incompleteness* result *because* (i) PA is consistent, (ii) since Gödel's constructed formula $\varphi$ denies the existence of a natural number with a particular property, $\varphi$ is determinately true or false, (iii) $\varphi$ is true *iff* $\varphi$ is undecidable in PA, (iv) $\varphi$ is undecidable in PA *iff* PA is consistent, and, therefore, *PA is incomplete* because (v) $\varphi$ is true *iff* PA is consistent (i.e., from (iii) and (iv)). Thus, given (i) and (v), $\varphi$ is true and undecidable in PA.

Much to the contrary, according to Wittgenstein, we cannot prove the incompleteness of PA because we cannot *prove* that no natural number has the property in question. The only sense in which $\varphi$ could be true is if we have a *proof* showing *how* it is the case that *any* natural number does not have the property in question. Only a proof by mathematical induction can establish that *any* natural number has a particular property (i.e., *not* that *all* naturals have a property). But according to Gödel's proof, such a proof is *impossible*.

The official story claims that although it is not possible to prove $\sim F(1), \sim F(2), \sim F(3)$, etc., if PA is consistent, there *exist* infinitely many facts *which are not knowable*. There *exist* mathematical facts such that 1 is not $F$ and 2 is not $F$, etc. We have, therefore, infinitely many *mathematical facts* and a general fact which are *unknowable*. And there are, allegedly, infinitely many propositions $\sim F(1), \sim F(2), \sim F(3)$, etc., which are provable, but *whose provability* is *not* provable. To use Hilbert's own 1931 words, "a science like mathematics must not rely upon faith, however strong that faith may be" (Hilbert, 1931, p. 268).

Part and parcel of Wittgenstein's syntactical conception of mathematics is that the very notion of incompleteness presupposes an unacceptable, sterile, and false Platonism. The claim that there exist causally inefficacious mathematical facts (and/or objects) which make mathematical propositions true and false is Platonism. The incompleteness *interpretation* of Gödel's *conditional independence proof* is intrinsically Platonistic. This interpretation *presupposes* Platonism—the proof does *not* force (or prove) a Platonistic distinction between syntax and semantics in mathematics. This Platonistic presupposition is notoriously problematic (e.g., the causation problem) and entirely unnecessary in light of the physical facts that constitute the *doing* of real mathematics by physical mathematicians using brain, pencil and paper.

## 8 Unanswerable Mathematical Questions?

Interestingly, one conclusion that could be drawn from this sketch is that *both* Wittgenstein and Hilbert can accept Gödel's syntactical, conditional proof without acquiescing to the conclusion that mathematics or number theory is incomplete. For Wittgenstein, as we have seen, this conclusion makes no sense. For Hilbert, however, who *gave* the conclusion its sense, he only need give up the demand that all axiomatic mathematical systems be Hilbert-complete. He is *not* compelled to capitulate to du Bois-Reymond on the mathematical front, since Gödel and others claim that there are *extra-systemic reasons* that *establish* that the Gödel-sentences are true, and that such reasoning *decides* such questions. On this, the received view, certain propositions are undecidable axiomatically and yet decidable extra-axiomatically (i.e., meta-mathematically). Thus, for a particular axiomatic system such as PA, if one is *given* a true but undecidable-in-PA Gödelian proposition, one can, *in principle*, establish that it is number-theoretically true. Indeed, this was Hilbert's almost immediate reaction when he proposed his $\omega$-rule in 1931. Hence, Gödelian propositions are not unanswerable mathematical questions and Hilbert is right in saying that in mathematics there is no *ignorabimus*.

For Wittgenstein, however, the most obvious conclusion to draw is that what poses as *a priori* intuition is nothing more that overt or covert Platonism. Only the vacuous claim that there exist mathematical realms about which we can *intuitively* and *knowingly* make clear and intelligible claims gives any credence to the belief that 'questions' such as the Continuum Hypothesis and GC speak about a clearly understood and clearly defined realm of entities. If we drop this unfounded belief, and moor our calculi in extra-systemic real-world applications, we will *leave* and *return* to domains of contingent propositions which will guide us in extending our calculi in various useful directions.

Like Hilbert, but for markedly different reasons, Wittgenstein is also not compelled to capitulate to du Bois-Reymond on the mathematical front, for, on Wittgenstein's view, the fact "that an equation is a rule of syntax... explain[s] why we cannot have questions in mathematics that are in principle unanswerable" (Wittgenstein, 1975, §121).

## References

Brouwer, L. E. J. (1907). *Over de Grondslagen der Wiskunde*. Amsterdam: Maas & van Suchtelen. (Doctoral Thesis. English translation *"On the Foundations of Mathematics"* in (Heyting, 1975, pp. 11–101).)

Brouwer, L. E. J. (1908). De onbetrouwbaarheid der logische principes. *Tijdschrift voor Wijsbegeerte*, 2, 152–158. (English translation "The Unreliability of the Logical Principles" in Heyting, 1975, pp. 107–111.)

Brouwer, L. E. J. (1929). Mathematics, science, and language. *Monatshefte für Mathematik, und Physik*, 36(1), 153–164. (Reprinted as "Mathematics, Science, and Language" in Mancosu, 1998, pp. 45–53.)

Dawkins, R. (2006). *The god delusion*. Boston: Houghton Mifflin Company.

Ewald, W. (Ed.). (1996). *From Kant to Hilbert: A sourcebook in the foundations of mathematics*. Oxford: Clarendon Press.

Feferman, A. B., & Feferman, S. (2004). *Alfred Tarski: Life and logic*. Cambridge: Cambridge University Press.

Feigl, H. (1981). The Wiener Kreis in America. In R. S. Cohen (Ed.), *Inquiries and provocations. Selected writings 1929-1974* (pp. 57–94). Dordrecht: D. Reidel.

Finch, H. (1977). *Wittgenstein—The later philosophy*. Atlantic Highlands, N.J.: Humanities Press.

Floyd, J., & Dreben, B. (1991). Tautology: How not to use a word. *Synthese*, 87, 23–49.

Gödel, K. (1931). Über formal unentscheidbare Sätze der Principia Mathematica und verwandter Systeme I. *Monatscheft für Mathematik und Physik*, *38*(1), 173–198. (English translation "On Formally Undecidable Propositions of Principia Mathematica and Related Systems I" in van Heijenoort (1967, pp. 596–616).)

Gödel, K. (1947). What is Cantor's continuum problem? *The American Mathematical Monthly*, *54*(9), 515–525.

Gödel, K. (1995). The present situation in the foundations of mathematics. In S. Feferman et al. (Ed.), *Collected works, Vol. III* (pp. 45–53). Oxford: Oxford University Press. (Original from 1933, previously unpublished.)

Gould, S. J. (1999). *Rocks of ages*. New York: The Ballantine Publishing Group.

Haeckel, E. (1899). *Die Welträthsel*. Bonn: Strauss. (English translation "The Riddle of the Universe at the Close of the 19th Century". Cambridge: Cambridge University Press, 2009.)

Heyting, A. (Ed.). (1975). *L.E.J. Brouwer: Collected works, Vol. I*. Amsterdam: North-Holland Publishing Company.

Hilbert, D. (1899). *Grundlagen der Geometrie*. Stuttgart: B.G. Teubner. (Reprinted as Hilbert, 1999.)

Hilbert, D. (1902). Mathematical problems. *Bulletin of the American Mathematical Society*, *8*(10), 437–479.

Hilbert, D. (1925). Über das Unendliche. *Mathematische Annalen*, *1926*(95), 161–190. (English translation "On the Infinite" in van Heijenoort, 1967, pp. 369–392.)

Hilbert, D. (1929). Probleme der Grundlegung der Mathematik. *Atti del Congresso internazionale dei matematici, Bologna 3-10 settembre 1928, (Bologna, 1929), Vol. 1*, 135–141. (English translation "Problems in the Grounding of Mathematics" in Mancosu, 1998.)

Hilbert, D. (1930). Naturerkennen und Logik. *Naturwissenschaften*, *28, November 1930, Vol. 18*, 959–963. (English translation "Logic and the Knowledge of Nature" in Ewald, 1996, pp. 1157–1165.)

Hilbert, D. (1931). Die Grundlegung der elementaren Zahlenlehre. *Mathematische Annalen*, *104 (1931)*, 485–494. (English translation "The Grounding of Elementary Number Theory" in Mancosu, 1998, pp. 266–273.)

Hilbert, D. (1999). *The foundations of geometry* (L. Unger, Trans.). La Salle, Illinois: Open Court.

Mancosu, P. (1998). *From Brouwer to Hilbert: The debate on the founda-*

*tions of mathematics in the 1920s.* Oxford: Oxford University Press.
Marion, M. (2003). Wittgenstein and Brouwer. *Synthese, 137*, 103–127.
Menger, K. (1994). *Reminiscences of the Vienna Circle and the Mathematical Colloquium* (L. Golland, B. McGuinness, & A. Sklar, Eds.). Dordrecht: Kluwer.
Moore, G. (1955). Wittgenstein's lectures in 1930-33, Part III. *Mind, 64*, 1–27.
Rodych, V. (1999). Wittgenstein's inversion of Gödel's theorem. *Erkenntnis, 51*(2/3), 173–206.
Rodych, V. (2002). Wittgenstein on Gödel: The newly published remarks. *Erkenntnis, 56*(3), 379–397.
Rodych, V. (2003). Misunderstanding Gödel: New arguments about Wittgenstein and New remarks by Wittgenstein. *Dialectica, 57*(3), 279–313.
Rodych, V. (2006). Who is Wittgenstein's worst enemy?: Steiner on Wittgenstein on Gödel. *Logique et Analyse, 49*(193), 55–84.
Skolem, T. (1923). Einige Bemerkungen zur axiomatischen Begründung der Mengenlehre. In *Matematikerkongressen i helsingfors den 4-7 juli 1922, den femte skandinaviska matematikerkongressen, redogörelse* (pp. 217–232). Helsinki: Akademiska Bokhandeln. (Reprinted in Skolem, 1970, pp. 137–152. Quotation translated in Wang, 1996, p. 131.)
Skolem, T. (1929). Über die Grundlagendiskussionen in der Mathematik. In *Proceedings of the 7th scand. math. congress* (pp. 3–21). (Reprinted in Skolem, 1970, pp. 207–225. Quotation translated in Wang, 1996, p. 121.)
Skolem, T. (1970). *Selected works in logic* (J. Fenstad, Ed.). Oslo: Universitetsforlaget.
Steiner, M. (2001). Wittgenstein as his own worst enemy: The case of Gödel's theorem. *Philosophia Mathematica, 9*(3), 257–279.
van Dalen, D. (2005). *Mystic, geometer, and intuitionist: The life of L.E.J. Brouwer: Hope and disillusion Vol. II.* Oxford: Clarendon Press.
van Heijenoort, J. (Ed.). (1967). *From Frege to Gödel: A sourcebook in mathematical logic.* Cambridge, Mass.: Harvard University Press.
Wang, H. (1987). *Reflections on Kurt Gödel.* Cambridge, Mass.: M.I.T. Press.
Wang, H. (1996). Skolem and Gödel. *Nordic Journal of Philosophical Logic, 1*(2), 119–132.
Wittgenstein, L. (1921). *Logisch-Philosophische Abhandlung.* An-

nalen der Naturphilosophie. (English translation "Tractatus Logico-Philosophicus", London: Routledge and Kegan Paul, 1961; translated by D.F. Pears and B.F. McGuinness.)

Wittgenstein, L. (1975). *Philosophical remarks* (R. Rhees, Ed. & R. Hargreaves & R. White, Trans.). Oxford: Basil Blackwell.

Wittgenstein, L. (1978). *Remarks on the foundations of mathematics* (G. von Wright, R. Rhees, & G. Anscombe, Eds. & G. Anscombe, Trans.). Oxford: Basil Blackwell.

Victor Rodych
Department of Philosophy
University of Lethbridge
Lethbridge, Alberta, T1K 3M4, Canada
e-mail: Rodych@uleth.ca

# From Pair Points to Pairs of Models

IGOR SEDLÁR[1]

**Abstract:** The paper offers an alternative interpretation of pair points, discussed in (Beall, Brady, Dunn, & al., 2012). A pair point is explicated as a single point viewed from two different perspectives. The interpretation serves as a motivation for formulating a semantics using pairs of Kripke models (pair models). It is demonstrated that, if certain conditions are fulfilled, pair models are validity-preserving copies of positive substructural models. This yields a general soundness and completeness result for a variety of (positive) substructural logics with respect to multimodal Kripke frames with binary accessibility relations.

**Keywords:** accessibility relations, pair points, substructural logics

## 1 Introduction

Giving a 'philosophical story' behind the formal semantics of a deductive system is far from being a 'pleasant extra'. If the semantics is supposed to model a fragment of natural language, for example, one has to explain the connection of the semantics with our 'pre-theoretical' intuitions concerning the fragment. What is the connection between boolean algebras and propositional connectives, such as 'and' or 'or'? What is the connection between 'It is necessary that' and Kripke models?

The more complicated the semantics is, the more pressing the *interpretation problem* becomes. One of its notorious instances is related to the *ternary semantics* of substructural logics:[2]

> [...] whereas the the binary relation invoked by Kripke in the semantics of modal logics has several philosophically interesting and revealing interpretations (as relative possibility, or as

---

[1] The author is greatly indebted to the audience at *Logica 2012* for their valuable comments, as well as to the organisers for such an enjoyable conference (again). This paper was written in the Department of Logic and Methodology of Sciences, Comenius University, as a part of the research project VEGA 1/0046/11, *Semantic models, their explanatory power and applications*.

[2] However, many special cases of the ternary semantics have received plausible interpretations. These are in most cases related to the original motivations for introducing the respective deductive systems (Lambek calculus, Linear logic etc.). See (Restall, 2000), for example.

a temporal ordering, or as the relation of being-morally-ideal-from-the-point-of-view-of, or ... ), the ternary relation invoked by Routley and Meyer has no such standardly accepted interpretations/applications. (Beall et al., 2012, p. 597)

The problem is felt to be quite important especially if the ternary relation is to comply with rather complicated requirements as to its properties. This is the case of several *relevant* logics, for example.

Let us consider the 'ternary case' of the problem in more detail. It may be approached from at least two different angles. First, the problem might be *solved* by providing a philosophical interpretation of the ternary relation.[3] Second, it might be *avoided* by demonstrating that the ternary relation is not essential. Put differently, one can try to provide a semantics for substructural logics without the ternary relation (perhaps with binary relations only). Investigations along these lines have been outlined in (Dunn, 1976) and (Kurtonina, 1995, 1998) for example.[4]

The recent paper (Beall et al., 2012) contains many interesting suggestions as well. It is suggested there that the ternary relation might be seen as a binary relation between points and ordered pairs of points – *pair points*.[5] This suggestion shall be our staring point. Section 2 explains the idea of pair points in more detail. Moreover, it offers an alternative interpretation: a pair point is seen as a *single* point viewed from two different *perspectives*. Section 3 expands on the idea in a more formal manner. The notion of a pair model representation of a (positive) substructural model is introduced and it is explained that the representations correspond to *pairs* of multimodal Kripke models. The section also discusses a slightly more abstract notion: a pair frame representation of a substructural frame. Truth conditions related to substructural connectives are given and the central result of the paper is established: Every (positive) substructural logic characterised by a class of substructural frames is sound and complete with respect to a class of multimodal Kripke frames with binary accessibility relations. The section concludes by explaining that the result opens a new perspective on substructural connectives. Section 4 concludes the paper and points out the most important open problems.

---

[3] See (Restall, 1995) and (Mares, 2004), for example.

[4] There is also the failed attempt (Vasyukov, 1986), see (Dunn, 1987). A corrected version appeared as (Vasyukov, 1994).

[5] This observation builds upon a remark by Routley and Meyer (Meyer & Routley, 1973). It follows the standard mathematical practice of defining $\langle x, y, z \rangle$ as $\langle x, \langle y, z \rangle \rangle$. See also (Beall, 2009).

## 2 Pair Points

This section explains the idea of pair points in more detail. We begin by outlining the motivation for considering pair points in (Beall et al., 2012) as well as the related technical details (2.1). Then we sketch the original interpretation of pair points and offer a new one (2.2). The new interpretation is the background of the technical work of Section 3.

### 2.1 Pair Points and Counterexamples

Beall et al. (2012) observe that the conditional defined in terms of the ternary relation seemingly bucks the 'no counterexample' interpretation. According to the interpretation, $A \to B$ is true at a point $x$ iff there is no relevant counterexample $y$ such that $A$ is true at $y$, whilst $B$ is false at $y$. Quite so! We can have $x \not\Vdash A \to B$ because there are *distinct* points $y, z$ such that $Rxyz$, $y \Vdash A$, and $z \not\Vdash B$. In general, no *single* counterexample point that makes $A$ true and $B$ false is required to make a conditional $A \to B$ false at a given point.

Nonetheless, there is a simple way out: extend the notion of a point to include ordered pairs of 'old' points as well. More precisely, *pair points* $\langle xy \rangle$ may be introduced, where $x, y$ are 'old' points. Now the ternary $R$ may be rephrased as a binary relation between points and pair points: $Rxyz$ simply means $Rx\langle yz \rangle$. Consequently, pair points may serve as counterexamples for false conditionals.

However, two points need clarification. First, the truth conditions of $A \to B$ should be restated in terms of pair points. This, in turn, requires a clear notion of truth and falsity at pair points. Beall et al. (2012) proceed as follows. Truth $\models_T$ and falsity $\models_F$ at pair points are defined in terms of the original forcing relation $\Vdash$:

- $\langle xy \rangle \models_T A$ iff $x \Vdash A$.

- $\langle xy \rangle \models_F A$ iff $y \not\Vdash A$.

Hence, $\langle xy \rangle \models_T A$ is in general consistent with $\langle xy \rangle \models_F A$.[6]

The rephrased truth condition of $A \to B$ runs as follows:

- $x \models_T A \to B$ iff there is no $\langle yz \rangle$ such that $Rx\langle yz \rangle$ and $\langle yz \rangle \models_T A$, but $\langle yz \rangle \models_F B$.

---

[6]However, this is not the case in *half points* $\langle xx \rangle$, which can be identified with the 'old' points of the model. (A pair point $\langle xy \rangle$ where $x \neq y$ is called a *duo point*.)

- $x \models_F A \to B$ iff $x \not\models_T A \to B$ ($x$ may be seen as $\langle xx \rangle$).

It is plain that the conditional is now susceptible to the 'no counterexample' interpretation. If $A \to B$ is false at $x$, then there is a counterexample $\langle yz \rangle$, such that $Rx\langle yz \rangle$, $\langle yz \rangle \models_T A$, but $\langle yz \rangle \models_F B$.

### 2.2 Pair Points and Perspectives

The original interpretation of pair points appeals to the theory of situated inference, see (Mares, 2004). Half points (or, equivalently, the 'old' points) are seen as situations (see Barwise & Perry, 1983) and duo points as *information links*. Examples of information links include 'laws of nature, conventions, and any information that gives us a license to make inferences' (Beall et al., 2012, p. 602).

But other interpretations are possible as well. Let us begin by considering the following example. Let $T$ denote the statement that a given person $a$ is tall. The two statements $T$ and $\neg T$ may be seen as describing two different states of affairs ($a$ as an adult and $a$ as a child, for example). But it is also possible to see these statements as describing the *same* state of affairs from *two distinct points of view*. For example, $a$ might by tall with respect to her family, but not tall with respect to the local basketball team.

The same observation applies to pair points. One approach is to see pair points as pairs of two (possibly distinct) situations. But there is also a quite different angle: the pair point $\langle xy \rangle$ may be seen as a *single* situation viewed from *two distinct perspectives*. For example, $x, y$ may correspond to belief sets of two distinct agents in the situation $\langle xy \rangle$ (their different 'opinions' about the situation).

There is a simple way to model the difference between the above approaches. Let us, for the time being, consider only points and valuations without any reference to accessibility relations. Take a pair $(P, V)$, where $P$ is a set of points and $V$ is a valuation on $P$. The *pair point model* built on $(P, V)$ is the pair $(P^2, V)$. If we decided to extend $(P, V)$ by a ternary relation, the relation could be simulated as a binary relation on $P^2$ in the pair point model.

But the pair point model itself might be simulated without explicitly using pairs of points of the original $(P, V)$. The key is to replace the somewhat vague 'perspectives' by *valuations*. The *pair representation* of $(P^2, V)$ is a triple $(P', V_1, V_2)$, such that there is a bijection $\sigma$ from $P'$ to $P^2$ and $V_1, V_2$ are two valuations. Now $w \in P'$ might be seen as a *representation* of $\sigma(w)$ iff the following holds: $w_i \in V(p)$ iff $w \in V_i(p)$ for $i \in \{1, 2\}$, where $w_i$ is the $i$-th member of $\sigma(w)$. Put differently, the 'binary nature' of $\langle xy \rangle \in P^2$ might be simulated by using two valuations $V_1, V_2$ 'operating on' points that are not explicit pairs. For a simple example, see 2.2.

Now, obviously, we have an interesting twist in the story: pair representations might be seen as *pairs* of models of the original form: $(P', V_1)$ and $(P', V_2)$. Thus, pair point models correspond to pairs of 'simple models'. The following section develops this observation and applies it to substructural models.

## 3 Pair Representations

This section develops the interpretation of pair points, outlined in Section 2. The well-known definitions of (positive) substructural models and frames are outlined. In addition, pair models and pair frames are defined. The truth conditions of substructural formulas are adapted to pair models. Pair model and pair frame representations of substructural models and frames respectively are defined (3.1). Then, the main result of the paper is established (3.2).

### 3.1 The Basic Definitions

We shall be working with a positive substructural language, built upon a countable set of propositional variables $\Phi$. Formulas of the language contain $\Phi$ and are built by applying the binary operators $\wedge$ (extensional conjunction), $\vee$ (extensional disjunction), $\rightarrow$ (implication), $\circ$ (intensional conjunction, fusion) and $\leftarrow$ (converse implication).[7]

---

[7] However, our results hold for languages using any subset of these operators.

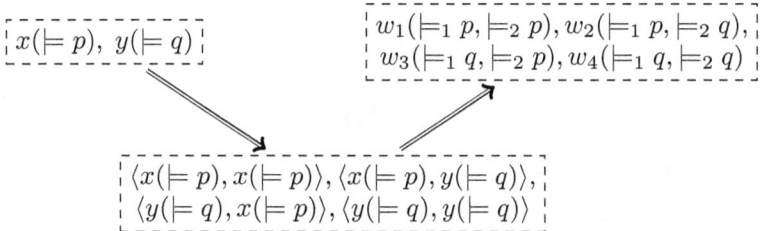

Figure 1: Pair points and their simulation by means of two valuations.

Validity shall be defined for *consecutions* of the form

$$X \vdash A$$

where $A$ is a formula and $X$ is a *structure* built from formulas by means of applying the binary operations ',' (comma) and ';' (semicolon). For more details, see (Restall, 2000).

**Definition 1** (Restall, 2000)  *A (positive) substructural frame is a triple*

$$\mathfrak{F} = (P, \sqsubseteq, R)$$

*where $P$ is a non-empty set (of 'points'), $\sqsubseteq$ is a partial order on $P$ and $R \subseteq P^3$.*

*A (positive) substructural model built on a frame $\mathfrak{F}$ is a couple*

$$\mathfrak{M} = (\mathfrak{F}, \Vdash)$$

*where $\Vdash$ is a forcing relation between points and members of $\Phi$ such that $x \sqsubseteq y$ and $x \Vdash p$ imply $y \Vdash p$ for all $p \in \Phi$.*

*The forcing relation can be extended to every formula and structure in a familiar way (see Restall, 2000). A consecution $X \vdash A$ is valid in $\mathfrak{M}$ iff $x \Vdash X$ implies $x \Vdash A$ for all $x \in \mathfrak{M}$, i.e. all $x$ in $P$, where $P$ belongs to $\mathfrak{M}$ (notation: $X \vdash_{\mathfrak{M}} A$). $X \vdash A$ is valid in a frame $\mathfrak{F}$ iff it is valid in every $\mathfrak{M}$ built on $\mathfrak{F}$ (notation: $X \vdash_{\mathfrak{F}} A$). If $\mathfrak{C}$ is a class of substructural frames, then $X \vdash A$ is valid in $\mathfrak{C}$ iff it is valid in every $\mathfrak{F} \in \mathfrak{C}$ (notation: $X \vdash_{\mathfrak{C}} A$).*

*A (positive) substructural logic L is* characterised *by a class of frames $\mathfrak{C}$ iff the following holds: $X \vdash_{\mathfrak{C}} A$ iff $X \vdash A$ is provable in L.*

**Definition 2**  *A pair frame is a triple*

$$\mathbf{F} = (W, R_0, \{R_j^i\})_{i,j \in \{1,2\}}$$

*where $W$ is a non-empty set and $R_0$, $R_j^i$ are binary relations on $W$.*

*A pair model built on a frame $\mathbf{F}$ is a couple*

$$\mathbf{M} = (\mathbf{F}, \{V_i\})_{i \in \{1,2\}}$$

*such that $\mathbf{F}$ is a pair frame and $V_i$ are valuations, i.e. functions from $\Phi$ to subsets of $W$.*

A pair model is a set of points together with five binary relations and two valuations. Hence, a pair model M might be seen as a pair of multimodal Kripke models $\langle \mathbf{M}_1, \mathbf{M}_2 \rangle$. For example:

- $\mathbf{M}_1 = (W, R_0, R_1^1, R_2^1, V_1)$
- $\mathbf{M}_2 = (W, R_1^2, R_2^2, V_2)$

A pair frame is simply a multimodal Kripke frame. (But, on the other hand, it may be seen as a pair of Kripke frames as well.)

**Definition 3** *The valuations $V_1, V_2$ give rise to two truth relations $\models_1$ and $\models_2$. These are defined as follows ($i \in \{1,2\}$):*

- $(\mathbf{M}, w) \models_i p$ *iff* $w \in V_i(p)$.
- $(\mathbf{M}, w) \models_i A \wedge B$ *iff* $(\mathbf{M}, w) \models_i A$ *and* $(\mathbf{M}, w) \models_i B$.
- $(\mathbf{M}, w) \models_i A \vee B$ *iff* $(\mathbf{M}, w) \models_i A$ *or* $(\mathbf{M}, w) \models_i B$.
- $(\mathbf{M}, w) \models_i A \to B$ *iff* $R_1^i wv$, $(\mathbf{M}, v) \models_2 A$ *and* $R_0 vu$ *imply* $(\mathbf{M}, u) \models_1 B$, *for all $v, u \in W$.*
- $(\mathbf{M}, w) \models_i A \circ B$ *iff there are $v, u \in W$ such that $R_1^i wv$, $R_0 uv$, $(\mathbf{M}, u) \models_1 A$, and $(\mathbf{M}, u) \models_2 B$.*
- $(\mathbf{M}, w) \models_i B \leftarrow A$ *iff* $R_2^i wv$, $(\mathbf{M}, v) \models_1 A$ *and* $R_0 vu$ *imply* $(\mathbf{M}, u) \models_1 B$, *for all $v, u \in W$.*

*These conditions are extended to structures similarly as it is done for substructural models (see Restall, 2000). Hence, ';' mimics '$\circ$' while ',' mimics '$\wedge$'. A consecution $X \vdash A$ is* valid in $\mathbf{M}$ *iff $(\mathbf{M}, w) \models_1 X$ implies $(\mathbf{M}, w) \models_1 A$, for all $x \in W$ (notation: $X \vdash_\mathbf{M} A$).*

*A consecution $X \vdash A$ is* valid in a frame $\mathbf{F}$ *iff it is valid in every $\mathbf{M}$ built on $\mathbf{F}$ (notation: $X \vdash_\mathbf{F} A$). If $\mathbf{C}$ is a class of pair frames, then $X \vdash A$ is* valid in $\mathbf{C}$ *iff it is valid in every $\mathbf{F} \in \mathbf{C}$ (notation: $X \vdash_\mathbf{C} A$).*

*A (positive) substructural logic $L$ is* characterised *by a class of pair frames $\mathbf{C}$ iff the following holds: $X \vdash_\mathbf{C} A$ iff $X \vdash A$ is provable in $L$.*

**Definition 4** *A pair frame $\mathbf{F}$ is a* pair frame representation *(a p.f.r.) of a substructural frame $\mathfrak{F}$ iff there is a bijection $\sigma : W \to P^2$ such that ($w_i$ denotes the $i$-th member of the couple $\sigma(w)$):*

- $R_0 wv$ *iff* $Rw_1 w_2 v_1$ *and*
- $R_j^i wv$ *iff* $w_i = v_j$.

*A pair model* $\mathbf{M} = (\mathbf{F}, \{V_i\})_{i,j \in \{1,2\}}$ *is a* pair model representation *(a p.m.r.) of a substructural model* $\mathfrak{M} = (\mathfrak{F}, \Vdash)$ *iff:*

- $\mathbf{F}$ *is a p.f.r. of* $\mathfrak{F}$ *and*
- $w \in V_i(p)$ *iff* $w_i \Vdash p$.

*If* $\mathfrak{S}$ *is a substructural model, frame or a class of frames, then the respective class of pair representations is denoted* $\mathsf{Rep}(\mathfrak{S})$.

Note that $\mathsf{Rep}(\mathfrak{F})$, $\mathsf{Rep}(\mathfrak{M})$ are non-empty, for every $\mathfrak{F}, \mathfrak{M}$. In addition, observe that if $\mathbf{M} = (\mathbf{F}, \{V_i\})_{i,j \in \{1,2\}}$ and $\mathbf{F} \in \mathsf{Rep}(\mathfrak{F})$, then $\mathbf{M} \in \mathsf{Rep}(\mathfrak{M})$ for some $\mathfrak{M}$ built on $\mathfrak{F}$.

Hence, a *p.m.r.* of $\mathfrak{M}$ represents the information contained in $\mathfrak{M}$ by means of a pair of multimodal Kripke models. The first step is to rephrase the substructural model by means of pair points: the 'new' points are pairs $\langle xy \rangle$ of the 'old' points. From this point of view, the ternary $R$ may be replaced by a binary $R'$ such that $R' \langle xy \rangle \langle zz' \rangle$ iff $Rxyz$. Now the points $w \in \mathbf{M}$ can be seen as representations of the pairs $\langle xy \rangle$, if there is a bijection that preserves their properties and their 'position' among other pairs. Of course, this position and their properties are given by i) the relation $R$, ii) the inner structure of the pairs (e.g. the information that the second point in $\langle xy \rangle$ is the same as the first point in $\langle yz \rangle$ is quite important), iii) the valuation. Now $R_0$ 'models' the binary $R'$ and, therefore, the ternary $R$. The relations $R_j^i$ are there to 'keep track' of the inner structure of the modelled pairs. The notation itself suggests that $R_j^i wv$ means that the $i$-th member of the pair represented by $w$ (i.e. of $\sigma(w)$) is identical with the $j$-th member of the pair represented by $v$ (i.e. of $\sigma(v)$). Last but not least, the pair of valuation reflect the 'binary nature' of pair points (the principle has been explained in Section 2.2).

It is now obvious that the seemingly awkward truth conditions for formulas containing $\rightarrow, \circ, \leftarrow$ are 'mere' reformulations of the corresponding truth conditions in substructural models. As we shall see in the following section, this yields a general result about the characterisation of (positive) substructural logics by means of pair frames.

### 3.2 Pair Frames and Substructural Logics

**Lemma 5** *Let* $\mathbf{M}$ *be a p.m.r. of* $\mathfrak{M}$. *Then*

$$(\mathbf{M}, w) \models_i A \text{ iff } (\mathfrak{M}, w_i) \Vdash A$$

*for every bijection $\sigma$ with the properties specified in Definition 4. The same holds for structures.*

*Proof.* The basic case $A = p$ holds by Definition 4. The cases $A = B \wedge C$, $A = B \vee C$ are trivial.

Next, assume that $w \not\models_i B \to C$. This means that there are $v, u$ such that $R_1^i wv$, $R_0 vu$, $v \models_2 B$, but $u \not\models_1 C$. By the induction hypothesis and by Definition 4, this amounts to $w_i = v_1$, $Rv_1v_2u_1$, $v_2 \Vdash B$, but $u_1 \not\Vdash C$. However, this holds iff $w_i \not\Vdash B \to C$. This completes the proof for case $A = B \to C$. The cases $A = B \circ C$ and $A = C \leftarrow B$ are proved similarly. The cases for structures are very similar to cases for $\wedge$ and $\circ$. $\square$

**Lemma 6**

a) $X \vdash_\mathfrak{M} A$ iff $X \vdash_{\mathsf{Rep}(\mathfrak{M})} A$.

b) $X \vdash_\mathfrak{F} A$ iff $X \vdash_{\mathsf{Rep}(\mathfrak{F})} A$.

c) $X \vdash_\mathfrak{C} A$ iff $X \vdash_{\mathsf{Rep}(\mathfrak{C})} A$.

*Proof.* a) Assume that $X \not\vdash_\mathfrak{M} A$. There is a point $x \in \mathfrak{M}$ such that $(\mathfrak{M}, x) \Vdash X$, but $(\mathfrak{M}, x) \not\Vdash A$. Now let $\mathbf{M}$ be any $p.m.r.$ of $\mathfrak{M}$ and let $\sigma(w) = \langle xy \rangle$ for some $y \in \mathfrak{M}$. By Lemma 5, $(\mathbf{M}, w) \models_1 X$, but $(\mathbf{M}, w) \not\models_1 A$. Consequently, $X \not\vdash_{\mathsf{Rep}(\mathfrak{M})} A$.

Now assume that $X \not\vdash_{\mathsf{Rep}(\mathfrak{M})} A$, i.e. there is a $p.m.r.$ $\mathbf{M}$ of $\mathfrak{M}$ such that $(\mathbf{M}, w) \models_1 X$, but $(\mathbf{M}, w) \not\models_1 A$ for some $w \in \mathbf{M}$. By Lemma 5, $(\mathfrak{M}, w_1) \Vdash X$, but $(\mathfrak{M}, w_1) \not\Vdash A$. Hence, $X \not\vdash_\mathfrak{M} A$.

b) Assume that $X \not\vdash_\mathfrak{F} A$. There is a model $\mathfrak{M} = (\mathfrak{F}, \Vdash)$ and a point $x$ such that $(\mathfrak{M}, x) \Vdash X$, but $(\mathfrak{M}, x) \not\Vdash A$. By Lemma 5, every $\mathbf{M} \in \mathsf{Rep}(\mathfrak{M})$ invalidates $X \vdash A$ as well. By Definition 4, if such $\mathbf{M}$ is built on a frame $\mathbf{F}$, then $\mathbf{F} \in \mathsf{Rep}(\mathfrak{F})$. Hence $X \not\vdash_{\mathsf{Rep}(\mathfrak{F})} A$.

Assume $X \not\vdash_{\mathsf{Rep}(\mathfrak{F})} A$. There is a model $\mathbf{M}$ built on a frame $\mathbf{F}$ and a point $w$ such that $\mathbf{F} \in \mathsf{Rep}(\mathfrak{F})$, $(\mathbf{M}, w) \models_1 X$, but $(\mathbf{M}, w) \not\models_1 A$. By the remark following Definition 4, there is substructural $\mathfrak{M} = (\mathfrak{F}, \Vdash)$, such that $\mathbf{M} \in \mathsf{Rep}(\mathfrak{M})$. By Lemma 5, $X \not\vdash_\mathfrak{M} A$. Consequently $X \not\vdash_\mathfrak{F} A$.

c) is an immediate consequence of b). $\square$

**Theorem 7 (General Pair-Frame Theorem)** *If a (positive) substructural logic $L$ is characterised by a class of substructural frames $\mathfrak{C}_L$, then it is characterised by the class of pair frames $\mathsf{Rep}(\mathfrak{C}_L)$.*

*Proof.* Follows from Lemma 6 c). $\square$

Hence, many well-known positive substructural logics are characterised by multimodal Kripke frames.

However, the truth conditions of formulas containing $\to, \circ, \leftarrow$ are not ordinary modal clauses. Note again that pair models $\mathbf{M}$ can be seen as pairs of Kripke models $\langle \mathbf{M}_1, \mathbf{M}_2 \rangle$. Moreover, observe that, for example, the condition for $\models_1 A \to B$ refers also to $\models_2$ (similarly for $\circ, \leftarrow$). This shows that, in general, formulas with $\to, \circ, \leftarrow$ 'operate' *between* the models $\mathbf{M}_1$ and $\mathbf{M}_2$ in the pair model (or 'pair of models') $\mathbf{M}$.[8]

This gives us the following 'hierarchy of operators': boolean operators operate 'within' points in models; modal operators operate 'between' points, but always 'within' models; substructural operators operate between points and between models.

## 4 Conclusion

As we have noted in the Introduction, the interpretation problem of the ternary substructural semantics can be avoided by providing a *binary* semantics for substructural logics. Such a semantics (for positive logics) is provided in the above sections. However, many issues are left open.

First, there is an interpretation problem with respect to our semantics as well. The semantics might be seen as rather complicated, hence a 'philosophical story' behind it has to be provided. This will be done in an expanded version of this paper. This story shall be an epistemic one, explaining the two valuations and five relations in terms of information sharing within an ordered pair of agents. Consequently, the substructural connectives shall receive an epistemic reading.

Second, the treatment of negation is an important issue, though not necessarily one very hard to work out. This shall be dealt with in an expanded version of the paper as well.

Third, one might long for an 'independent description' of pair model representations and of specific classes of pair frame representations (corresponding to specific substructural logics). In other words, we should like to identify a set of conditions $X$ such that a pair model satisfies $X$ iff it is a pair model simulation of a substructural model. In addition, we should like to identify a set of conditions $X_{\mathfrak{E}}$ such that a pair frame satisfies $X_{\mathfrak{E}}$

---

[8] A familiar example of similar 'inter-model' truth conditions are the conditions for *public announcement* formulas $[A]B$ in public announcement logic. See (van Ditmarsch, van der Hoek, & Kooi, 2008).

iff it is a pair frame representation of a substructural frame $\mathfrak{F} \in \mathfrak{C}$. These investigations are also left for another occasion.

## References

Barwise, J., & Perry, J. (1983). *Situations and Attitudes*. MIT Press.
Beall, J. (2009). *The Spandrels of Truth*. Clarendon Press.
Beall, J., Brady, R., Dunn, J., & al. (2012). On the Ternary Relation and Conditionality. *Journal of Philosophical Logic, 41*(3), 595–612.
Dunn, J. (1976). A Kripke-style semantics for R-Mingle using a binary accessibility relation. *Studia Logica, 35*, 163–172.
Dunn, J. (1987). Incompleteness of the bibinary semantics for $r$. *The Bulletin of the Section of Logic, 16*, 107–109.
Kurtonina, N. (1995). *Frames and Labels: A Modal Analysis of Categorial Inference*. Unpublished doctoral dissertation, Utrecht University.
Kurtonina, N. (1998). Categorial inference and modal logic. *Journal of Logic, Language and Information, 7*(4), 399-411.
Mares, E. D. (2004). *Relevant Logic: A Philosophical Interpretation*. Cambridge University Press.
Meyer, R. K., & Routley, R. (1973). Classical relevant logics II. *Studia Logica, 33*, 183–194.
Restall, G. (1995). Information flow and relevant logics. In J. Seligman & D. Westerståhl (Eds.), *Logic, language and computation: The 1994 moraga proceedings* (pp. 463–477). CSLI Publications.
Restall, G. (2000). *An Introduction to Substructural Logics*. Routledge.
van Ditmarsch, H., van der Hoek, W., & Kooi, B. (2008). *Dynamic Epistemic Logic*. Springer.
Vasyukov, V. (1986). The bibinary semantics for $r$ and $Ł_{\aleph_0}$. *The Bulletin of the Section of Logic, 15*, 109–114.
Vasyukov, V. (1994). From ternary to tetrary? *The Bulletin of the Section of Logic, 23*, 163–167.

Igor Sedlár
Department of Logic and Methodology of Sciences
Comenius University
Gondova 2, 814 99 Bratislava, Slovakia
e-mail: sedlar@fphil.uniba.sk
URL: https://sites.google.com/site/sedlarsite/

# Towards Being

HARTLEY SLATER

**Abstract:** This paper is a response to Graham Priest's recent book 'Towards Non-being' (Priest, 2005). The specific discussion of Priest's work on Intensional Logic forms the third section. Before that is a presentation of a contrary account derived directly from Hilbert's Epsilon Calculus (which Priest also uses), and the paper starts with a consideration of one other matter that reflects on Priest's work: the issue of whether there are 'true contradictions'.

**Keywords:** contradictions, the epsilon calculus, individuals, individuating properties

## 1 Avoiding True Contradictions

It is because people, under the influence of the current formal tradition forget the indexicality in language that they get into such problems as the Liar Paradox. For one frequently hears arguments about sentences like 'This is not true' generating contradictions, even 'true contradictions'. Surely, if that sentence is true then, because of what it says then it must be not true, and so not true, absolutely. But also, surely, then it is not not true, i.e. true, again because of what it says. The reasoning, of course, is fallacious because it hides the presumption that the verbal item 'this' in the sentence has a specific reference that the sentence alone does not carry. The intention, or attempted intention, is to refer, with 'this' not to the sentence itself but to some proposition it is taken to express given the meaning then associated with the demonstrative. But what proposition is that? Until the referent of the subject term is provided there is no proposition expressed, and yet a referent would be needed prior to the intended proposition being identified via what was predicated. The situation is no better if one gives, say, numbers to sentences, even Gödel numbers, since they too could have many referents. If sentence 1 is 'sentence 1 is not true' then the referent of the 'sentence 1' in quotes is still to be given, and must not be presumed to be the same as the referent of the 'sentence 1' outside of the quotes. The reference of a term like 'sentence 1' is dependent on the context of its use, and thus it is not confined to be the reference given to it in the present context. Because of this the disquotation required in Tarski's T-scheme is not guaranteed, invalidating it as a principle.

More generally, how does the sentential tradition get into its difficulty with the Liar Paradox? It does so because we can easily construct a sentence J such that J = 'sentence J is not true', i.e. '¬TJ'. And Tarski's Truth Scheme, if applied within the same language, is: T'p' ≡ p. So that leads to the well-known contradiction via the series of equivalences: ¬TJ ≡ ¬T'¬TJ' ≡ ¬¬TJ ≡ TJ. Suppose, however, we use Horwich's Equivalence Scheme instead, namely: T[p] ≡ p, where '[p]' is 'that p'. Then, obviously, we do not get a contradiction. We can say that ¬TJ, since now sentences do not have any truth-value. But ¬TJ ≡ ¬T'¬TJ', and not ¬T[¬TJ] (≡ TJ). Certainly if '¬TJ'= [¬TJ] then there would be a contradiction, but that would involve equating a mentioned sentence with a 'that'-clause. Of course, if the sentence '¬TJ' was *non-indexical*, then the proposition [¬TJ] would be the only thing that could be stated when using the sentence, and there would be the same paradox with sentence J* = 'the proposition stated by J* is not true'. For then, by the Equivalence Scheme, [the proposition stated by J* is not true] is true ≡ the proposition stated by J* is not true, while the proposition stated by J* is [the proposition stated by J* is not true]. So all this means that '¬TJ' cannot be unambiguous.

But also the Liar Paradox is now seen to arise only with Tarski's T-scheme. For Horwich's 'That p is true', while it is of the subject-predicate form, is equivalent to 'It is true that p', which is of an operator form. And 'It is true that p' involves the operator 'it is true that' which is the null or identity operator in the modal system KT. But one modal fact in KT is that if L*p ≡ p then it is not the case that p ≡ ¬L*p, since KT is consistent. And there is a quite general fact about all the 'L' operators in KT: ¬L(p ≡ ¬Lp).

The point about the indexicality of '¬TJ' also applies to 'the proposition stated by J* is not true', and arises with Horwich's problematic sentence 'THE PROPOSITION FORMULATED IN CAPITAL LETTERS IS NOT TRUE'. Horwich thinks that this shows the Liar Paradox re-appears with propositional truth (Horwich, 1998, pp. 41–42). Haack's similar case concerns a sentence numbered 1 = 'the statement made by the sentence numbered 1 is not true' (Haack, 1978, p. 150). Don't we get with these that G = [¬TG] (for some proposition G), and so that ¬TG ≡ ¬T[¬TG] ≡ TG? No. Anyone who thinks so is forgetting the possibility that, outside of a context, no definite proposition is made in such cases, as no proposition would be made if the referring phrase ostensibly referring to a proposition or statement were replaced by a demonstrative such as 'this'. For then it would follow that the sentence on its own would not state anything definite to be not true. In fact it follows by *Reductio* from the supposed contradic-

Towards Being 173

tion above, that the definite description 'the proposition stated by J*' must be non-attributive, i.e. Millian. Likewise with Horwich's and Haack's referential phrases.

## 2  The Epsilon Account of Individuals

Turning now to the subjects of propositions, in Russell's theory of descriptions there are, it will be remembered, three clauses with 'The King of France is bald'. These are 'there is a king of France', 'there is only one king of France' and 'he is bald'. Russell used an iota term to symbolise the definite description, but of course this is not an individual symbol since 'The King of France is bald' on Russell's account does not have the elementary form 'Bx'. Russell hypothesised that, in addition to the linguistic expressions gaining formalisations by means of his iota terms, there was another, quite distinct class of expressions, which would take the place of the variable in such forms as 'Bx'. He suggested that demonstratives might be in this class, but he could give no further formal expression to them. The epsilon theory of descriptions that settles the question was discussed in the first edition of Hughes and Cresswell's classic introductory text on Modal Logic. It originated with Routley, Meyer and Goddard, who, in their work on intensional contexts, made an explicit identification of definite descriptions with epsilon terms: The King of France = $\epsilon x(Kx.(y)(Ky \supset y = x))$, (Goddard & Routley, 1973, p. 558; Hughes & Cresswell, 1968, p. 203; Routley, 1977; Routley, Meyer, & Goddard, 1974).

Which theorems in the epsilon calculus are behind this kind of identification? The standard epsilon calculus contains the axiom '$(\exists x)Fx \supset F\epsilon x Fx$' (Leisenring, 1969; Meyer Viol, 1998), from which one can naturally obtain the equivalence between the two sides. This makes existence a matter of an individual 'living up to its name', i.e. actually having the properties inscribed in its referential description. There is then one theorem in particular which demonstrates strikingly the relation between Russell's attributive, and Donnellan's 'purely referential' understanding of referential terms. For, following on from the above equivalence it can be proved that

(1) $(\exists x)(Kx \,\&\, (y)(Ky \supset y = x) \,\&\, Bx)$,

is logically equivalent to

(2) $(\exists x)(Kx \,\&\, (y)(Ky \supset y = x)) \,\&\, Ba$,

where a = $\epsilon x(Kx \,\&\, (y)(Ky \supset y = x))$, this being the epsilon term arising from the first conjunct in (2). The first expression, as we have seen, encapsulates Russell's Theory of Descriptions, in connection with 'The K is B'; it involves the explicit assertion of the first two clauses, to do with the existence and uniqueness of a K. Since Donnellan, however Donnellan (1966), we have realized that there are no preconditions on the introduction of 'the K' as an individual term. So 'The K is B', with 'The K' an individual term, may always be given a truth value, even if, sometimes, that truth value is merely an arbitrarily chosen one. For 'Ba' properly formalises 'The K is B', since the cross-reference in (2) means that it reads 'There is a single K. It is B', and the descriptive replacement for the E-type pronoun 'it', there, is 'The K'. On this basis (1) is better read 'A sole K exists and is B' rather than 'The K is B'. If the description in 'a' is non-attributive, i.e. if the first two clauses of Russell's account are not both true, and a sole K does not exist, then the referent of 'the K' is simply up to the speaker to nominate.

What epsilon terms formalise more generally are demonstratives, in line with Russell's identifying such as 'logically proper names', in his lectures on Logical Atomism. One major thinker who has missed this is Graham Priest, who took from Kneebone the idea that epsilon terms formalise indefinite descriptions, making the epsilon symbol replace 'an'. But if there is no uniqueness clause requiring a reading in terms of 'the', then '$\epsilon xFx$' is best read 'that F', as arises, for instance, when reading '$(\exists x)Fx \,\&\, G\epsilon xFx$' as 'There is an F. That F is G'. The epsilon terms in these kinds of contexts replace E-type pronouns in ordinary speech. These are anaphoric pronouns, and so point to other elements in the preceding discourse. Thus in this last case '$\epsilon xFx$' replaces 'It', while Routley, Meyer and Goddard's definite description 'the king of France' replaces 'He' in the context given before. In Donnelan's historic case with the phrase 'the man with martini in his glass' used referringly, the speaker(s) are just selecting a referent for the epsilon expression '$\epsilon x(Mx.Gx)$' when $\neg(\exists x)(Mx.Gx)$, in line with the general semantics for epsilon terms. Thus if $(\exists x)Fx$ then the referent of '$\epsilon xFx$' is to be selected from amongst the Fs, while if $\neg(\exists x)Fx$ then it can be selected arbitrarily from the world at large.

How can something be the one and only K 'if there is no such thing', i.e. if there is nothing with the character inscribed in the term? That is where a second, and even more important theorem in the epsilon calculus is required:

$$(Ka \,\&\, (y)(Ky \supset y = a)) \supset [a = \epsilon x(Kx \,\&\, (y)(Ky \supset y = x))].$$

For the singular thing is that this entailment cannot be reversed, so there is a difference between the left hand side and the right hand side, e.g. between something being alone king of France, and that thing being the one and only king of France. The difference is not available in Russell's logic, since only possession of the property can be formalised there. In fact Russell confused the two forms, since possession of an identifying property he formalised using the identity sign, viz 'a = $\iota$xKx', making it appear that some, maybe even all identities are contingent. But all proper identities are necessary, and it is merely associated identifying properties that are contingent. That means that in all possible worlds there is the same domain of discourse, but the individuals in that domain may change their properties, and even their individuating properties, from one world to the next.

One further consequence is that while individuals have eternal existence, they must be separated from any entities that merely have 'existence' in this world, or some other. For what, in connection with *individuals*, has 'existence' just in this world, or just in some other, making them 'physical objects', and 'fictions', respectively, are not the individuals themselves, but their *identifying properties*. To highlight the difference even more, we can say that *Aristotelian Realism* holds for such physical objects/fictions, whereas *Platonic Realism* holds for the associated individuals. The metaphysical point is important in connection with the shift from a mathematical to a literary point of view required for a proper understanding of intensions. For the difference between the two kinds of Realism is illustrated most clearly in the previous epsilon calculus theorem, which shows that

$$(\exists x)(Kx \,\&\, (y)(Ky \supset y = x) \,\&\, Bx)$$

(i.e. 'A sole king of France exists and is bald') is equivalent to

$$(\exists x)(Kx \,\&\, (y)(Ky \supset y = x)) \,\&\, B\,\epsilon x(Kx \,\&\, (y)(Ky \supset y = x))$$

(i.e. 'A sole king of France exists. He is bald'). The first conjunct in the second expression is about certain identifying properties being instantiated. That is what must hold for a sole king of France to exist (contingently).

The second conjunct there, however, is about a certain eternally existing individual: one that is a sole king of France if there is such a thing, but which still exists even if there is no such thing.

But not only through presenting Russell's formula as a conjunction do we enable a separation to be made between a true or false assertion about this world, namely the first conjunct delimiting existence and uniqueness conditions, and a further assertion, in the second conjunct, which is made about its subject independently of whether the first conjunct is true or false, and so about something that exists eternally. For the same point is central to understanding how such eternally real objects are accessed, which is a seemingly perennial difficulty with Platonic entities. Paradigmatically the situation is represented again in the epsilon variant to Russell's analysis of 'The king of France is bald'. For the first conjunct in the second expression above is itself equivalent to a conjunction:

$$K \, \epsilon x(Kx \,\&\, (y)(Ky \supset y = x)) \,\&\, (y)(Ky \supset y = \epsilon x(Kx \,\&\, (z)(Kz \supset z = x))).$$

So access to the individual $\epsilon x(Kx \,\&\, (z)(Kz \supset z=x))$, i.e. the king of France, is provided entirely by means of the linguistic act of supposing there is a sole king of France, and through its then being invariably possible to cross-refer to the same individual from within further assertions. Eternal objects, in this way, are simply subjects of discourse.

In particular one must remember, at this point, that ordinary proper names are disguised descriptions, and so one must distinguish the chosen individual from the associated property. For instance, we can say that p = $\epsilon x$(x is called 'Pegasus'), just as r = $\epsilon x$(x is called 'Russell'). So each of p and r exist, even though $\neg(\exists x)$(x is called 'Pegasus'), while $(\exists x)$(x is called 'Russell') etc.

## 3  Priest's Account of Individuals

Graham Priest has recently given an account of 'intentional' predicates and operators. There are a number of things common between our two accounts, which I shall first list; but mostly there are very wide divergences. Priest uses the epsilon calculus as I do. He says (Priest, 2011, p. 243): 'If something satisfies the condition A(x) in the actual world then $\epsilon xA(x)$ refers to one such (contextually determined) thing. Hence $A(\epsilon xA(x))$ is true. If nothing satisfies this condition then [$\epsilon xA(x)$] refers to some other, contextually determined object.' As a result he has a fair understanding of the required account of the round square $\epsilon x(Rx.Sx)$. He says (Priest, 2011, p. 243):

> Assuming the actual world is consistent, then nothing satisfies the condition 'x is round and square'. The description, then, will refer to some contextually determined object that does not satisfy this condition. What properties this object has depends, of course, on what object actually is picked out in the context. The fact that there is no determinate answer to the question of what properties it has is hardly objectionable, therefore any more than any other case of contextual determinacy.

Nevertheless he gets into difficulty with, amongst other things, what he calls 'non-existents.' He says (Priest, 2011, pp. 238, 243):

> Every world contains the same domain of objects, D. At each world an object may or may not exist. Thus there is a monadic existence predicate whose extension at a world is the set of thing[s] that exist there. So suppose that a term denotes a nonexistent object. What more can be said about its properties? For a start, it cannot have existence entailing properties, by definition. So it cannot be on top of the Berkeley Clock Tower, and I cannot kick it.

But the round square, in the above sense, *can* be on top of Berkeley Clock Tower, and one *might* kick it, if $\epsilon x(Rx.Sx)$ was chosen carefully enough. So Priest's account clearly needs to be amended, and in fact the necessary revision of it is very substantial indeed.

What, in summary, is right and what is wrong with Priest's account in comparison with the above? Well, his first error is that he defines propositions in terms of possible worlds rather than in terms of translations. This derives from the over-riding expectation in his culture that there be a mathematical analysis of logic, rather than a linguistic one. But then he includes 'impossible worlds' as well as possible ones, and propositions that may be both true and false, in tune with his well-known views about paraconsistency. This derives from his attachment to Tarski's theory of truth, and his lack of attention to the crucial differences between this and Horwich's theory. As well, Priest's account includes a view of individuals such that they may or may not exist; and he has contingent identities in addition to necessary ones. The point is not that Ockham's Razor is sufficient to eliminate all these unneeded beasts; it is the much more substantive one that these monstrosities are based on confusions, and are therefore incorrect. They are fictions in a surreal fantasy story we have been reading, and we must now return to the actual world, and regain the use of our senses. In particular we have seen that on the proper view of truth, there are no true contradictions. In addition we have seen that the same individuals exist in all worlds, making, at least, Priest's common domain D correct, but also making all identities between individuals necessary. Contingent 'identities' can only arise using something like Russell's iota term symbolism, in which an individuating description is masquerading as an individual term. With that understood then what has contingent existence are some properties, not any individual in which those properties are instantiated.

Priest shows he has some grasp of the required difference between individuals and their individuating descriptions. But he is held up through his belief that existence is to be formulated as a predicate of individuals, making some but not all individuals 'exist' at a world. So he misses the fact that contingent existence is a matter of the instantiation of properties, and the related fact that somethings looking like identities between individual terms are instead talking about the properties of an individual. He says (Priest, 2011, p. 239):

> *Prima facie*, it would appear that Oedipus desired Jocasta, but did not desire his mother, even though Jocasta is his mother. This is not a tough problem, however. Oedipus *did* desire his mother. He just did not realize that Jocasta was his mother. Of course he realized that Jocasta was Jocasta. Hence substitutivity of identicals, we may suppose, holds within the scope of intentional predicates.

He therefore thinks that substitutivity of identicals, while it holds for intensional predicates like 'desire' breaks down with intensional operators like 'realise that'. What Oedipus did not realize, however, was not an identity between individuals, but the property of an individual. What he failed to realise was that $j=\iota x Mxo$, not that $j=\epsilon x Mxo$, where 'Mxy' is 'x is mother of y' (cf. Slater, 1992). He realised that $j=j$, and so realised that $j=\epsilon x Mxo$, because in fact $j=\epsilon x Mxo$, and substitutivity of identicals holds even in operator constructions. But he could realize that $j=\epsilon x Mxo$ without realizing that $j=\iota x Mxo$, i.e. that $(y)(Myo \equiv y=j)$, either through thinking that he had a mother who was not Jocasta, or that $\neg(\exists x) Mxo$.

More particularly Priest misses just how true it can be, in real life, that we pity Anna Karenina herself (cf. Slater, 1987). For while he wants to say that intensional relations can exist with things that do not exist, and allows that we can say 'I pity Anna Karenina', he does so while realising that 'Sherlock Holmes', and so presumably 'Anna Karenina', is merely a disguised description (in Russell's terminology). So the relation of pity we have is a more general one to something with appropriate characteristics, showing that pitying Anna naturally involves pitying other things with similar descriptions. But that allows we can pity even her, if the associated epsilon term, encapsulating the full description, is chosen appropriately. In the actual world the referent of this epsilon term '$\epsilon x A x$' is strictly arbitrary since no-one has all the required characteristics, i.e. $\neg(\exists x)Ax$. But still we might locate that referent as a specific person amongst Anna's counterparts; for instance an actor on the stage who we see as her.

These points also relate centrally to the debate over emotional responses to fictions. For one must remember that many readers, film viewers, etc. believe the fictions they are involved with are actual, or are similar enough to actual events, or are quite likely true etc., and it is on that kind of basis that understandable human relations might be developed with what would otherwise be just arbitrary entities. Formally some such further presumption must be added before we can have an emotional relation with fictions, otherwise the much-discussed paradox of fiction would arise (Radford & Weston, 1975). For it is not the case that such full emotional relations are experienced without belief (even with non-fictions). Fictions, on their own, are 'dead', and so the possibility always is there of viewing them entirely 'aesthetically', i.e. as (Platonic) real objects without any belief in their (Aristotelian) existence.

## References

Donnellan, K. (1966). Reference and definite descriptions. *Philosophical Review*, 75, 281–304.
Goddard, L., & Routley, R. (1973). *The logic of significance and context*. Aberdeen: Scottish Academic Press.
Haack, S. (1978). *Philosophy of Logics*. Cambridge: C.U.P.
Horwich, P. (1998). *Truth*. Oxford: Clarendon.

Hughes, G., & Cresswell, M. (1968). *An introduction to modal logic.* London: Methuen.
Leisenring, A. (1969). *Mathematical Logic and Hilbert's Epsilon Symbol.* London: Macdonald.
Meyer Viol, W. (1998). *Instantial Logic.* Amsterdam: ILLC.
Priest, G. (2005). *Towards Non-Being: The Logic and Metaphysics of Intentionality.* Oxford: O.U.P.
Priest, G. (2011). Against against non-being. *The Review of Symbolic Logic, 4,* 237–253.
Radford, C., & Weston, M. (1975). How can we be moved by the fate of Anna Karenina? *Proceedings of the Aristotelian Society, 49,* 67–93.
Routley, R. (1977). Choice and descriptions in enriched intensional languages 1, 2, 3. In E. Morscher, J. Czermak, & P. Weingartner (Eds.), *Problems in logic and ontology.* Graz: Akademische Druck- und Velagsanstalt.
Routley, R., Meyer, R., & Goddard, L. (1974). Choice and descriptions in enriched intensional languages. *Journal of Philosophical Logic, 3,* 291–316.
Slater, B. H. (1987). Fictions. *British Journal of Aesthetics, 27*(2), 145–155.
Slater, B. H. (1992). Descriptive Opacity. *Philosophical Studies, 66,* 167–181.

Hartley Slater
Philosophy M207, University of Western Australia
35 Stirling Highway, Crawley, WA 6009, Australia
e-mail: hartley.slater@uwa.edu.au
URL: http://msc.uwa.edu.au/philosophy/about/staff/hartley_slater/

# Flow of Time in BST/BCont Models and Related Semantical Observations

PETR ŠVARNÝ[1]

**Abstract:** First the Branching Space-time and Branching Continuations models are briefly presented. We compare their properties with the traditional definition of a Flow of Time from physics and we point out the difficulties this definition in relativistic time. A preliminary solution how to implement the Flow of Time in the given models is then proposed.

**Keywords:** branching, temporal logics, semantics, continuations

## 1 Introduction

This paper explores the semantical issues and possibilities of two temporal branching structures, Branching Space-time (BST) and Branching Continuations (BCont). In order for the models to be of any use, they should include our intuitive notions connected to time. One of these notions is the flow of time (FoT). Our attempts to incorporate FoT into the BST/BCont models also yield interesting results about the semantical nature of these structures. We mention the basic idea of BST/BCont in the first section. The second section is devoted to the presentation of the traditional FoT definition and the problems it faces in relativistic physics, and in the end we give a possible solution how to incorporate it into BST/BCont.

## 2 BST and BCont

The two formal approaches to space-time studied in this article are Branching space-time, first introduced by N. Belnap in (Belnap, 1992), and Branching Continuations, formed by T. Placek in (Placek, 2011). The notion B-models is used in this article in cases where we refer to both models. We present only basic rudiments of both systems here.

---

[1] I must give much credit and thanks to Thomas Müller and Dennis Dieks for their comments and suggestions to the article. The article would remain quite enigmatic and redundantly complicated without their contribution. Výstup projektu Vnitřních grantů 2012 Filozofické fakulty UK. Made with the support of the Internal grant VG107 of the Faculty of Arts, Charles University, 2012.

## 2.1 BST

The basis of a BST model is $\langle W, \leq \rangle$. In other words, we have a set of point events, called *Our World* and denoted $W$, which is partially ordered by $\leq$. This set represents all the possibilities and options of the world. We can find in $W$ a particular kind of sets of events called histories. These sets are the maximal directed subsets of $W$. The directedness property describes sets that have for every two of their members a common upper bound. These histories represent possible scenarios of events. It was T. Müller, who prepared a version of BST called Minkowski branching structure (MBS) where histories are isomorphic to Minkowski space-time. We leave any further motivations, details, or explanations to the reader. They can all be found in (Belnap, 2003) or (Placek & Belnap, 2011).

## 2.2 BCont

BCont was introduced as an attempt to create a version of BST capable of working with spatiotemporal holes (such as singularities)(Placek, 2011). The basic change lies in the construction of histories. Instead of a BST history, we content ourselves with any nonempty consistent subset of $W$. The question of consistency is founded on the connectivity of two points in the model via a snake-link. A snake-link is a point-event path, a succession of point-events where every two point-events are comparable by $\leq$. These replacements for histories are then called large events or just l-events. We recommend the reader to pay attention to the proposed semantics of BCont—the so called Branching Time+*Instants*-like models[2] in the original paper (Placek, 2011). These semantics take BCont closer to the older idea of Branching Time, i.e. a usual tree structure, and thus force l-events to be chains.

# 3 Flow of Time

A common perception of time is seeing it as a succession of events, in other words, a flow of time (FoT). We aim to incorporate this notion in some way into the B-models. We now address the question of how the formal notion of a flow of time in logic differs from the flow of time as it is understood in physics and present the basic requirements for a FoT.

---

[2] Abbreviated as BT+*I*.

We follow two introductory articles to temporal logics (Venema, 2001) and (Hodkinson & Reynolds, 2006). In these articles, we can find the term 'flow of time' described as a pair $(T, \prec)$, where $T$ is a non-empty set of time points and $\prec$ is a irreflexive and transitive binary relation on $T$, i.e. a strict partial order. Actually, FoT means any kind of temporal structure in usual temporal logics. It can mean also branching or circular models. However, we would like to take our motivation from physics. We could name this view as the 'classical' one, as Belnap says in the founding article of BST (Belnap, 2003): a 'classical view' on time understands it as a succession of infinite Euclidean spaces.

A 'modern' view on FoT can be found in (Dieks, 1988). Dieks describes how we can assign a different now-point to every worldline and generate a partial order based on the linear order of now-points on the worldlines. This process has one important restriction: no now-point should lie in the interior of the conjunction of past lightcones of other now-points. This construction of FoT can be used, if one stays careful, even in special relativity. Let us sum up what this notion entails according to Dieks:

**Definition 1** (Generalized FoT)   *A **generalized flow of time** fulfils the following:*

1. *worldlines with a linear order of now-points*

2. *ontological definiteness of past and present*

3. *a relation (or set) of now-points on worldlines respecting ontological definiteness*

At this point, we can turn to the B-models to see how generalized flow of time can be formalized in those models.

## 4   Does time flow in B-models?

To sum up our point of departure, we can see that if we intend to study FoT in B-models we must ask ourselves what kind of FoT we want to study. As we already know, studying the pure logical meaning of the term is useless as it covers also branching structures. However, we can study a different type of FoT. In our case it is the generalized FoT. We follow the idea of D. Dieks and we leave behind the notion from temporal logic. A basic formalization of a generalized FoT could be $(wl, N, W, \leq)$, where $wl$ are worldlines, $N$ is the set of now-points, $W$ are point-events with $\leq$ being the partial ordering of these points. In addition, the set $N$ follows the rule of ontological definiteness. As we might already see from this draft formalization, our primary view is closer to McTaggart's B-series and relies on an ordering of events. We investigate how a generalized FoT can relate to B-models and how could we incorporate the basic formalization into B-models. We address the models in the following order: BT+$I$-like models, BCont, and BST.

### 4.1 BT+$I$-like models of BCont

The reason why we start with BT+$I$-like models of BCont should not be a surprising fact as their basis, the original Prior/Thomason models, were close to the intuitive notion of a flow of time. Let us have a model $\langle W, \leq, S \rangle$. A member of $S$, an instant, is in this case a spatio-temporal location as defined in Def. 2. These are ordered by the relation $\precsim$. Thanks to the Fact 15 and Fact 16 from (Placek, 2011), we know that this ordering is dense, partial and even linear. Our first observation is thus straightforward and simple. A BT+$I$-like model comes very close to fulfil our formal demands for a flow of time. We just need to find a way to put that into a formal statement. We list some definitions for reference[3]:

**Definition 2** (S-t locations (Placek, 2011)) *We say that a model $\langle W, \leq \rangle$ of BCont has* spatio-temporal locations *iff there is a partition $S$ of $W$ such that*

1. *For each l-event $A$ and each $s \in S$, the intersection $A \cap s$ contains at most one element;*

2. *$S$ respects the ordering $\leq$, that is, for all l-events $A$, $B$, and all $s_1, s_2 \in S$, if all the intersections $A \cap s_1, A \cap s_2, B \cap s_1$ and $B \cap s_2$ are nonempty, and $A \cap s_1 = A \cap s_2$, then $B \cap s_1 = B \cap s_2$;*

3. *similarly for the strict ordering $<$;*

4. *if $e_1 \leq e_2 \leq e_3$, then for every l-event $A$ such that $s(e_1) \cap A \neq \emptyset$ and $s(e_3) \cap A \neq \emptyset$, there is an l-event $A'$ such that $A \subseteq A'$ and $s(e_2) \cap A' \neq \emptyset$, where $s(e_i)$ stands for a (unique) $s \in S$ such that $e_i \in s$;*

---

[3] All referenced material has a bibliography reference number in its title. If the referenced definition is altered in some way, this reference number is accompanied by an apostrophe.

5. *if L is a chain of choice events in $\langle W, \leq \rangle$ upper bounded by $e_0$ and such that $\exists s \in S \forall x \in L \exists e \in W : (x < e \wedge s(e) = s)$, then $\exists e^* \left( e^* \in \bigcap_{x \in L} \Pi_x \langle e_0 \rangle \wedge s(e^*) = s \right)$.*

**Definition 3** (Ordering of s-t locations (Placek, 2011))  *For $s_1, s_2 \in S$, $S$ being the set of s-t locations, let $s_1 \precsim s_2$ iff $\exists e_1, e_2 (e_1 \in s_1 \wedge e_2 \in s_2 \wedge e_1 \leq e_2)$.*

BT+*I*-like models were used by Placek as a neat way to introduce semantics and valuation on the BCont framework (or at least on a special kind of BCont). It seems suitable to use the motivation of BT+*I*-like models and treat FoT as a question of valuation. It suffices to use Def. 23 from (Placek, 2011):

**Definition 4** (Point fulfils formula (Placek, 2011))  *For given $e_C, e/A$ and the model $\mathfrak{M} = \langle \mathfrak{G}, \mathcal{I} \rangle$, then:*

1. *if $\psi \in Atoms$: $\mathfrak{M}, e_C, e/A \Vdash \psi$ iff $e \in \mathcal{I}(\phi)$[4];*

2. *if $\psi$ is $\neg \varphi$ : $\mathfrak{M}, e_C, e/A \Vdash \psi$ iff it is not the case that $\mathfrak{M}, e_C, e/A \Vdash \varphi$;*

3. *for $\wedge, \vee, \rightarrow$ also in the usual manner;*

4. *if $\psi$ is $F_x \varphi$ for $x > 0$ : $\mathfrak{M}, e_C, e/A \Vdash \psi$ iff there are $e' \in W$ and $e^* \in A$ such that $e' \leq e^*$ and $int(e', e, x)$, and $\mathfrak{M}, e_C, e'/A \Vdash \varphi$;*

5. *if $\psi$ is $P_x \varphi, x > 0$ : $\mathfrak{M}, e_C, e/A \Vdash \psi$ iff there is $e' \in W$ such that $e' \cup A \in l$-events and $int(e', e, x)$ and $\mathfrak{M}, e_C, e'/A \Vdash \varphi$;*

6. *if $\psi$ is $Sett : \varphi : \mathfrak{M}, e_C, e/A \Vdash \psi$ iff for every evaluation point $e/A'$ from fan $\mathcal{F}_{e/A}$ and $\mathfrak{M}, e_C, e/A' \Vdash \varphi$;*

7. *$Poss : \psi := \neg Sett : \neg \psi$;*

8. *if $\psi$ is $Now : \varphi : \mathfrak{M}, e_C, e/A \Vdash \psi$ iff there is $e' \in s(e_C)$ such that $e' \cup A \in l$-events and $\mathfrak{M}, e_C, e'/A \Vdash \varphi$.*

It is the last clause that should catch our attention. Observe that either $e' = e_C$ or the two events are inconsistent, in other words belong to different continuations of a choice point earlier than $e_C$.

---
[4]We use $\Vdash$ instead of Placek's $\approx$ purely for technical reasons, the meaning is the same.

**Lemma 5** *For given $e_C, e/A$ and the model $\mathfrak{M} = \langle \mathfrak{G}, \mathcal{I} \rangle$. If $\mathfrak{M}, e_C, e/A \Vdash Now : \varphi$ and thus there is a $e'$ such that $\mathfrak{M}, e_C, e'/A \Vdash \varphi$ then the point-event $e'$ mentioned in Def. 4 is either equal to $e_C$ or is inconsistent with it.*

*Proof.* If $e' = e_C$ then by definition $e' \in s(e_C)$. However, $e' \cup A \in$ l-events holds only if $A$ is consistent with $e_C$. If they are not consistent then $e_C$ cannot be $e'$ as the evaluation point needs compatibility of $e'$ and $A$. Therefore let us choose an $e'$ different from $e_C$. If it were consistent with $e_C$, $A$ would also need to be consistent with $e_C$. However, $s(e_C) \cap A'$ for any $A'$ has according to Def. 2 at most one member. If they are consistent, we can choose $A' = \{e'\} \cup \{e_C\} \cup A$ and thus only one possible outcome of the intersection, namely $e_C$. Hence supposing $e'$ and $e_C$ are consistent but different also leads to a contradiction. □

We can now introduce the basic idea of generalized FoT (gFoT). We understand gFoT as an ordering of sets of events deemed as contemporary to a given point-event $e$. The sets are ordered according to a worldline. We also need to follow the points mentioned in the Def. 1. The promise of the introduction of gFoT is that it could alleviate the work with the original BCont model. As Placek (2011) puts it:

> We take the Kripke/Prior/Thomason semantics for our reference theory, since it is relatively simple and we have some intuitions concerning tenses. We do not have comparable intuitions concerning relativistic notions, and for this reason it will not be revealing to take BST for our reference theory.

We could place weaker constrains on BCont models by using gFoT as a reference theory instead of BT and thus exploit much more from their potential. We first introduce the notions in the context of the familiar and easier BT+*I*-like models, later to be re-evaluated in pure BCont and also in BST models.

Some important choices have to be made at this point. Namely, how do we want to identify a given set of 'now'-points; do we want it to also reach different continuations or be only valid in one continuation? The point $e_C$ already serves in the original approach as a point of reference, thus there seems to be nothing wrong in using this point again. Addressing the second question, we saw in Def. 1 that there should be a difference between indefinite future and definite past. If we were to hold only onto one continuation, we could not make this distinction. On the contrary, making use of the available branching structure seems as a natural way to combine gFoT and branching, therefore now-points should also include possible continuations.

**Definition 6** (Setting of now-points)  $X_e$ *is a setting of now-points for the point-event $e$ iff $e \in W$ and for $X_e \subseteq W$ it holds that (1) $e \in X_e$, (2) $\forall x, y \in X_e : x \not< y \wedge y \not< x$, and (3)$\forall x \in X_e : x$ is consistent with $e$.*[5]

This is just a general linking of sets of now-points connected to a point of reference in a given continuation. We could also make the definition shorter by using the term space-like related (SLR) mentioned in (Placek, 2011). Two points are SLR if they are consistent but incomparable. Thus a setting of now-points is a set containing a reference point and points that are SLR with each other. We can also have a maximal setting of now-points, bearing in mind that being maximal does not yield uniqueness. This partially covers the need to avoid point-events from the past or future of a given point-event. However, as the construction of gFoT asks us for a worldline as well, we need to present a notion of worldline-like nature. We use a more general approach here than the BT+*I*-like models would need, where l-events are already chains. A more general definition, however, does not cause any harm. In fact, it can be used later for all BCont models.

**Definition 7** (Worldline)  *A set $Wl \subseteq W$ is a worldline iff $Wl$ is a chain. We denote $Wl_{(e)}$ a worldline containing the point-event $e$.*

We give the usual BCont chains a new name, although the two terms do not differ in any significant way. This definition might seem superfluous but we add it in order to comply with the original Def. 1 vocabulary.

**Definition 8** (Setting of now-points for a worldline)  *The $X_{Wl}$ is a setting of now-points for a worldline iff $Wl$ is a worldline and $X_{Wl} = \{X_e | e \in Wl \wedge \forall e, e' \in Wl(X_e \cap X_{e'} = \emptyset)\}$.*

In other words, the setting of a worldline is constituted of a disjunctive set of settings for the points of the worldline. One should keep in mind that this is a set of settings and thus a quite different notion from $X_e$. We also note some observations concerning the simplicity of these notions in the currently studied models.

**Lemma 9**  *In BT+I-like models, a setting of now-points for $e$ is $\{e\}$.*

*Proof.* The proof follows trivially from the given definitions. □

---

[5] Although the letter may seem as the letter 'X', it is the Greek Chi.

This observation is good to be kept in mind if one were to judge on the sense of introducing a new concept to BT+$I$-like models. The gFoT concept does not bring much new in these models. However, BT+$I$-like models also work with possible continuations. For this reason (and for future use in BCont) we take into account other continuations.

**Definition 10** (Setting of now-points in continuation) *The set $X_{e,A}$, the setting of now-points in continuation $A$ with respect to point-event $e$, is equal to the set $X_{e'}$ for some $e' \in s(e)$ such that $e'$ is consistent with $A$.*

**Definition 11** (Setting of now-points for a worldline in continuation) *The set $X_{Wl,A}$ is equal to the set $X_{Wl'}$, where $Wl'$ is a chain consistent with $A$, constituted of events $e' \in W$ such that $\forall e' \in Wl' \exists! e \in Wl : e' \in s(e)$ and $\forall e \in Wl \exists! e' \in Wl' : e \in s(e')$.*

**Lemma 12** *In BT+I-like models, if $e$ is inconsistent with $A$ then $X_{e,A}$ has a single point-event, namely $e' \in s(e)$, where $e'$ is consistent with $A$.*

*Proof.* The proof follows trivially from the given definitions and Lemma 5. □

There is a common idea to both definitions. One transforms the original points using spatiotemporal locations to the points consistent with the given l-event and constructs settings for these. With the exception of our approach there seems, in general, no reason to relate $X_{e,A}$ and $X_{e,A'}$, where $A$ is consistent with $e$ and $A'$ is not. One could, based on the physical motivations behind the whole project, for example imagine that continuation $A'$ leads the observer to some gravitational field and thus his setting of now-points should, quite naturally, be different from the observer in the continuation $A$.

Following our attempt to maintain a general approach, we also address the question of the interval function used in the definitions. The interval function was defined in the original paper (Placek, 2011) for some point events $e, e'$ and a coordinalization $\mathcal{X}$ as follows:

$$int(e, e', t) \text{ iff } \mathcal{X}(s(e')) - \mathcal{X}(s(e)) = t \tag{1.1}$$

As we can see, this definition explicitly builds on two notions that have a specific character in BT-like models. First, the ordering of $S$ in these models and second the coordinalization. The ordering is quite simple, coordinalization is therefore easily made and allows us to determine $t$ with mere subtraction. If we want to have a more general approach applicable also in the BCont models we can use worldlines.

**Definition 13** *For $e, e' \in W$, $A$ and l-event such that $e'$ is consistent with it, $Wl_{(e)}$ a worldline, and $\mathcal{X}$ being a coordinalization on the worldline: $int(e, e', Wl_{(e)}, t)$ is equivalent to:*

1. $\exists e'' : X_{e'',A} \in X_{Wl_{(e)},A} \wedge e' \in X_{e'',A}$
2. $\mathcal{X}(s(e'')) - \mathcal{X}(s(e)) = t$

We refer here to the same coordinalization as in the original article, i.e. an order preserving bijection $\mathcal{X}$ between $\langle S, \precsim \rangle$ mapped to the dense subset of $\mathfrak{R}$.

The general notion of interval does not, like the other altered definitions, change anything in the context of the simple BT+$I$ models. We can show this in a simple lemma.

**Lemma 14** *The interval function in BT+I models has the property $int(e, e', t) \equiv (\exists Wl_{(e)}) \, (int(e, e', Wl_{(e)}, t))$.*

*Proof.* The core idea that $t$ is equal to the difference of two points is the same in both definitions. The question is the identity of the two points. Let $int(e, e', t)$ be true. Also let $e, e'$ be consistent. Then there exists an l-event they belong to, let it be $A$. L-events are chains in BT+$I$, thus there exists a chain they both belong to. Let this chain be $Wl_{(e)}$. Because $e' \in A$ according to our assumptions, there exists $e''$ such that $X_{e'',A} \in X_{Wl_{(e)}} \wedge e' \in X_{e''}$. If we take into account Lemma 9 then there is only the option that $e' = e''$. Now if $e$ and $e'$ are inconsistent we have an l-event $A$ consistent with $e'$ (in the worst case it is the singleton of $e'$). Construct a chain $Wl_{(e)}$ such that $Wl_{(e)} \cap s(e') \neq \emptyset$. It follows from Def. 2 and from Fact 16 in (Placek, 2011)[6] that there is one element in BT+$I$ models in this intersection, let it be $e''$. Therefore $e'' \in s(e')$ and hence $\mathcal{X}(s(e'')) - \mathcal{X}(s(e)) = t$ is the same as $\mathcal{X}(s(e')) - \mathcal{X}(s(e)) = t$. Point-event $e''$ does verify all that we expect from it, by the definitions and Fact 16 it holds that $X_{e'',A} = \{e'\}$ and $X_{e'',A} \in X_{Wl_{(e)},A}$.

The other direction needs to be verified also. Let $\exists Wl(e) : int(e, e', Wl_{(e)}, t)$ be true and may the events be consistent. Then we ask: does it hold that $e' = e''$? And it does, as $X_{e''} \cap e' = e'$ in BT+$I$ models. If the two

---

[6]Linearity in BT+$I$ models.

events are inconsistent then obviously $e' \neq e''$ but for our proof it is enough to show that $s(e') = s(e'')$. This follows from $X_{e'',A} \cap e' = e'$ by definition and Lemma 12. □

At this point we can prepare a new evaluation. We return to the definition of how a point fulfils a formula but alter it to use the newly introduced notions.

**Definition 15** (Point fulfils formula—BT+$I$ and FoT)  *For given $e_C, e/A$ and the model $\mathfrak{M} = \langle \mathfrak{S}, \mathcal{I} \rangle$, the definition that a point fulfils a formula is the same as in Def. 4 with the exception of the following:*

- *if $\psi \in$ Atoms: $\mathfrak{M}, e_C, e/A, X_{Wl_{(e_C)},A} \Vdash \psi$ iff $e \in \mathcal{I}(\phi)$;*

- *if $\psi$ is $\neg\varphi$ : $\mathfrak{M}, e_C, e/A, X_{Wl_{(e_C)},A} \Vdash \psi$ iff it is not the case that $\mathfrak{M}, e_C, e/A \Vdash \varphi$;*

- *for $\wedge, \vee, \rightarrow$ also in the usual manner;*

- *if $\psi$ is $F_x\varphi$ for $x > 0$ : $\mathfrak{M}, e_C, e/A, X_{Wl_{(e_C)},A} \Vdash \psi$ iff there are $e' \in \bigcup X_{Wl_{(e_C)},A}$ and $e^* \in A$ such that $e' \leq e^*$ and $int(e, e', Wl_{(e)}, x)$, and $\mathfrak{M}, e_C, e'/A \Vdash \varphi$;*

- *if $\psi$ is $P_x\varphi$ for $x > 0$ : $\mathfrak{M}, e_C, e/A, X_{Wl_{(e_C)},A} \Vdash \psi$ iff there is $e' \in \bigcup X_{Wl_{(e_C)},A}$ such that $e' \cup A \in$ l-events and $int(e', e, Wl_{(e)}, x)$ and $\mathfrak{M}, e_C, e'/A \Vdash \varphi$;*

- *if $\psi$ is Sett : $\varphi$ : $\mathfrak{M}, e_C, e/A, X_{Wl_{(e_C)},A} \Vdash \psi$ iff for every evaluation point $e/A'$ from fan $\mathcal{F}_{e/A}$ : $\mathfrak{M}, e_C, e/A', X_{Wl_{(e_C)},A} \Vdash \varphi$;*

- *if $\psi$ is Now : $\varphi$ : $\mathfrak{M}, e_C, e/A, X_{Wl_{(e_C)},A} \Vdash \psi$ iff there is $e' \in X_{e_C,A}$ such that $e' \cup A \in$ l-events and $\mathfrak{M}, e_C, e'/A, X_{Wl_{(e_C)},A} \Vdash \varphi$.*

**Theorem 16** *Def. 4 is equivalent to Def. 15 in BT+I-like models.*

*Proof.* Some of the points from the Def. 4 were not changed in any significant way. Those points obviously hold as we only added a new notion to the right side of $\Vdash$ but it does not influence in any way the valuation in those cases.

The $F_x$ operator's equivalence: we need to prove that $e' = e'_{FoT}$, where $e' \in W$ comes from Def. 4 and $e'_{FoT} \in \bigcup X_{Wl_{(e_C)},A}$ is from Def. 15. We are using Lemma 14. Let us have $e'_{FoT} \in \bigcup X_{Wl_{(e_C)},A}$. It is a subset of $W$ and it also meets all the requirements of Def. 4 and hence it is an $e'$ point as required by it. According to Def. 13 and Def. 2 it even has to be equal to that point. For the other direction let us have $e' \in W$ that suits the Def. 4. Using downward directedness of BT+$I$ models for $e_{FoT}, e'$ there exists $e_0$ comparable to both events as both events are consistent with $A$, there can be an l-event $A'$ that both events belong to. The intersection of a s-t location and an l-event is only one point-event. Hence only one point-event fulfils $int(e, e', Wl_{(e)}, x)$ for a given $x$ and so $e' \in W = e'_{FoT}$. And a similar combination of BT+$I$ models' linearity and downward directedness leads to the equivalence of the P operator.

For the $Now$ operator, if we have $e'$ from Def. 15 then this exact $e'$ fulfils all that is needed for Def. 4 to work. For the other direction let us have $e'$ based on Def. 4. However, based on Lemma 12, $X_{e_C,A}$ is always a singleton. If $A$ is consistent with $e_C$ then it has $e_C$ as its single member and from the definition of setting of now-points (Def. 10) this single member must be $e'$, thus $e' = e_C$ and it fulfils all requirements of Def. 15. If $e_C$ is inconsistent with $A$ then the only change is that $e' \neq e_C$ but it is again a member of $X_{e_C,A}$. □

With this theorem, let us shift our attention to the other B-models as the results there promise to be less trivial. Our work in BT+$I$ served merely as a didactical or pragmatic training field before we enter the main battlefield of gFoT for which all the general approach was meant.

### 4.2 BCont

Our first attempt of use of the tested definitions is in the BCont structures without the limitations imposed on them by the BT-likeness. L-events do not have to be chains anymore. This is the main and most important change from the BT+$I$ framework. We use the pure BCont models with the addition of s-t locations as presented in (Placek, 2011). All the definitions stay unaltered in any way by the introduction of gFoT. However, we need to rethink the following definitions and theorems from (Placek, 2011):

- $int(e, e', t)$

- extensions of evaluation points

- fan of evaluation points

- point fulfils formula
- definite truth
- three values of definiteness

Some changes were already made in the previous section (the interval definition for example), other changes are inherited (e.g. extension of an evaluation point simply uses the new interval definition) and a few need to be completely redone with regards to BCont models. This section is devoted to the changes that weren't presented already in the BT+$I$ part or differ from those mentioned there. We also use the same structure and language as before. As a short reminder, the language $\mathcal{L}$ is made out of present tensed atomic formulas, classical logical connectives ($\wedge$ etc.), two metric temporal operators ($F_x$,$P_x$), two modal operators ($Sett$ :,$Poss$ :), and the operator $Now$. The semantical model is the same. The model for $\mathcal{L}$ is $\mathfrak{M} = \langle \mathfrak{G}, \mathfrak{I} \rangle$, where $\mathfrak{G} = \langle \mathcal{W}, \mathcal{X} \rangle$ is the structure and $\mathfrak{I} : Atoms \to \mathcal{P}(W)$ is an interpretation function. The structure is made out of a BCont model $\mathcal{W} = \langle W, \leq, S \rangle$ and $\mathcal{X}$, a real coordinalization of $S$.

**Definition 17** (Extensions of evaluation points (Placek, 2011))  *$e/A$ goes at least $x$-units-above $e$ $(0 \leq x)$ iff $\exists e_1 \in W \exists e_2 \in A \exists Wl \subseteq W : (e_1 \leq e_2 \wedge e_1 \in Wl \wedge e_2 \in Wl \wedge int(e_1, e_2, Wl, x)$*

*$e/A'$ is an $x$-units-above $e$ extension of $e/A$, $(0 \leq x)$ iff $A \subseteq A' \subseteq W$ and $e/A'$ goes at least $x$-units-above $e$.*

For the extension of evaluation points we only added the new definition of an interval. The fan of evaluation points and the so called instant-wise isomorphism can be left as they were in the original paper.

**Definition 18** (Fan of evaluation points (Placek, 2011))  *Two l-events $A_1, A_2$ of $W$ are isomorphic instant-wise iff $\forall e_1 \in A_1 \exists e_2 \in A_2$: $s(e_1) = s(e_2)$ and $\forall e_2 \in A_2 \exists e_1 \in A_1$: $s(e_1) = s(e_2)$;*

*A fan of evaluation points for $e/A$ is a set of evaluation points where $e/A' \in \mathcal{F}_{e/A}$ iff $e/A'$ is an evaluation point in $\mathfrak{G}$ and $A, A'$ are isomorphic instant-wise.*

We are ready to present how the fulfilment of a formula works in BCont with FoT. We can actually use the same definition as for BT+$I$, this would be the Definition 15.

**Definition 19** (Definite truth (Placek, 2011)') $\psi$ *is definite at* $\mathfrak{M}, e_C$, $e/A$, $X_{Wl_{(e_C)}}$, $A$ *iff there is an* $0 \leq x$ *such that for every x-units-above e extension of* $e/A'$ *of* $e/A$: $\mathfrak{M}, e_C, e/A, X_{Wl_{(e_C)}}, A \Vdash \psi$. *It is written as* $\mathfrak{M}, e_C, e/A, X_{Wl_{(e_C)}}, A \models \psi$.

$\psi$ *is indefinitely true at* $\mathfrak{M}, e_C, e/A, X_{Wl_{(e_C)}}, A$, *written* $\mathfrak{M}, e_C, e/A$, $X_{Wl_{(e_C)}}, A ?\models \psi$, *iff there is no* $0 \leq x$ *such that for every x-units-above e extension of* $e/A'$ *of* $e/A$: $\mathfrak{M}, e_C, e/A, X_{Wl_{(e_C)}}, A \Vdash \psi$ *or for every x-units-above e extension of* $e/A'$ *of* $e/A$: $\mathfrak{M}, e_C, e/A, X_{Wl_{(e_C)}}, A \Vdash \neg \psi$.

**Theorem 20** (Three options (Placek, 2011)) *For any formula* $\psi$ *and any evaluation point* $e/A$, *exactly one of the following three options must hold:*

$$e/A \models \psi \text{ or } e/A \models \neg\psi \text{ or } e/A ?\models \psi \qquad (1.2)$$

*Proof.* The proof is the same as used by Placek (2011). □

This presents our main topic with only one more addition. Placek (2011) has shown a few examples (he calls them puzzles) of how BT+$I$ valuation works. We use two of those examples, Peircean future and 'Was Einstein born a Nobel Prize winner' to demonstrate how the FoT valuation allows general BCont models to address similar topics.

*Peircean future*

Peircean approach means that a sentence in the future tense being true at $e$ means that it is true in every possible history to which $e$ belongs and thus fails to distinguish between what will happen and what will necessarily happen. As Placek's BT+$I$ models are able to distinguish between the two cases, so are we. We need to capture the difference between $F_x\psi$ and $Sett : F_x\psi$.

Similarly to the original paper, we can demonstrate this using a model $\mathfrak{M}$, where $\mathfrak{M}, e_C, e_C/A, X_{Wl_{(e_C)}}, A \models F_1\psi$ and it holds that $\mathfrak{M}, e_C, e_C/A$, $X_{Wl_{(e_C)}}, A \models \neg Sett : F_1\psi$.

An exemplar model is visualized on fig. 1 and it asks for some explanation. Although it does not incorporate all the notions from gFoT and BCont, it is already quite crowded.

This two dimensional model (one spatial dimension along the x axis, one time dimension along the y axis) is a subset of $W$. We see a worldline $Wl$, the curve going from the bottom to the top of the plane, and a point $e_C$ that belongs to the worldline. The point $e_1$ is a point such that

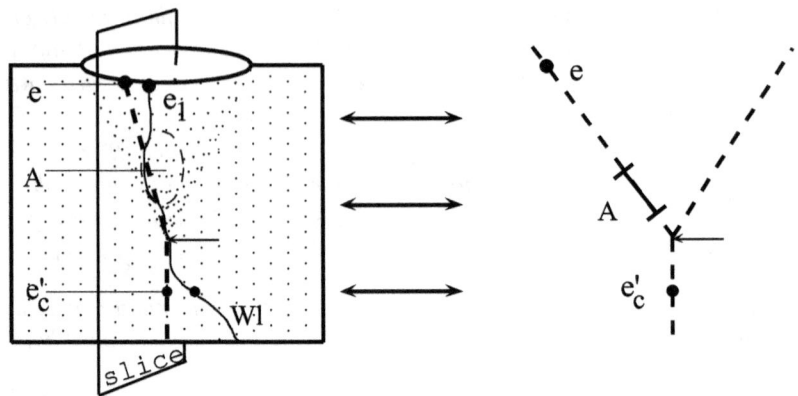

Figure 1: A 2D BCont model for Peircean future.

$int(e_C, e_1, Wl_{(e_C)}, 1)$ holds. The cone shows two possible continuations from a given choice event (an arrow points to it). The event $e$ represents then a member of $X_{e_1, A}$. If we wanted to see the two possible continuations, we could slice the model with a plane and get the image on the right side. There we can see the l-event $A$ (in the first picture it is the oval shape). The point $e'$ shows a member of $X_{e_1, A'}$. If at $e$ we fulfilled $\psi$ but at $e'$ $\neg\psi$ would hold, then we have a model that suits our purpose. $F_1\psi$ is fulfiled by every 1-units-above-$e_C$ extension of $e_C/A$, while every fan determined by each $x$-units-above-$e_C$ extension of $e_C/A$ has an element on the far side of the cone, where $\neg\psi$ is true at the given distance and setting of now-points.

This example also shows the usefulness of the gFoT notions. The worldline allows us to have some measure for intervals and the settings of now-points allow us to relate SLR points to each other and thus evaluate sentences using the operators of $\mathcal{L}$.

*Natural born Nobel Prize winner*

In the second example the sentence 'Einstein was born a Nobel Prize winner' is analysed. We assert it in the year 2012 and grant that Einstein might have failed to receive the Nobel Prize in 1921. The critical question to examine is the relation between $Sett : P_{100}F_9\psi$, called (S), and $P_{100}Sett : F_9\psi$, called (P), where $\psi$ stands for the sentence about Einstein. Placek (2011) has shown that (S) does not imply (P). The figure 2 helps us imagine the situation. The final figure is actually very similar to the figure shown in (Placek, 2011). Once again, we start out from the figure on the left with a similar description as before. The difference is now in the position of $e_C$ and the event we are looking for, namely the Nobel Prize ceremony of 1921 (marked N).

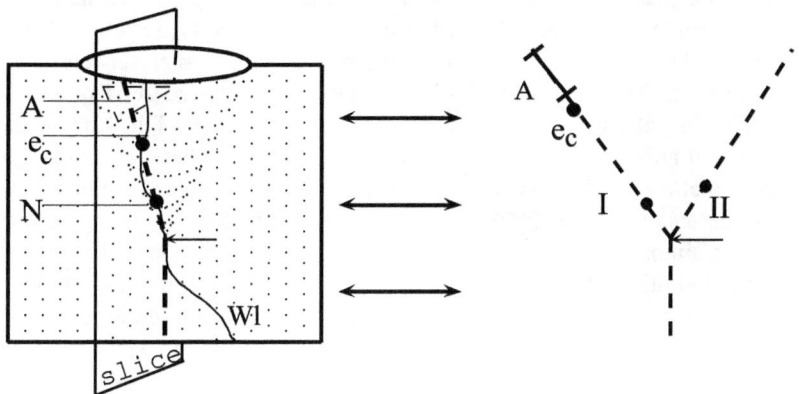

Figure 2: A 2D BCont model for $Sett : P_{100}F_9\psi \not\to P_{100}Sett : F_9\psi$.

For every $x$-units-above $e_C$ extension $\psi$ holds, it means that fans determined by any extension are made out of events that have N in their past. This means that (S) holds in the given model, on the contrary to (P) that does not hold. Our model is made in such a way that for every $e$ for which $int(e_C, e, Wl_{(e_C)}, 100)$ holds, this event does not have the choice event in its 'past' while it has either $N$ or $\neg N$ in its 'future'[7]. We need to say for every event $e$ as the settings of now-points might be different. For reasonable settings of now-points and a reasonable $Wl$ we would just take the event 100 units to the past on the worldline. Hence we are evaluating $Sett : F_9\psi$ in the event $e$. However, fans determined by any extension $e/A'$ from this point-event also have the elements from the possible continuation (II) where N did not occur and hence (P) does not hold because neither holds $\mathfrak{M}, e_C, e/A', X_{Wl_{(e_C)}, A'} \models Sett : F_9\psi$ nor does $\mathfrak{M}, e_C, e_C/A', X_{Wl_{(e_C)}, A'} \models P_{100}Sett : F_9\psi$.

---

[7] Neither of these was formally introduced, but are simple placeholders for the defined terms of 'strictly above' and 'strictly below'

### 4.3 BST

It was already shown by Placek (2011) that not every BCont model is a BST model. Hence we cannot simply take our results from Bcont and apply them to a supposed subset of BCont models because BST does not represent such a subset. We need to verify all the steps made with BCont models to make sure that the results are also valid in BST. However, the main goal of this article was achieved and these results are merely a sketch of how BST could be treated. The reason for interest in BST is its closer relation to special relativity and also its MBS interpretation. We tested gFoT on BCont models. In order to introduce gFoT and a valuation in BST it is necessary to have all notions we need for this task. We follow the basic definitions of BCont that are used in our gFoT approach and try to find BST equivalents for them(using Belnap, 2003 or Placek & Wroński, 2009). To prepare such list, one simply takes the Def. 15 and decomposes all its members to the crucial definitions for their existence. We do not list definitions that were introduced in this article because if we have the following notions, we can introduce for example settings of now-points in a similar way as we did here. In this way we get the following list:

- partially ordered model
- chains of events
- alternatives of events
- space-like related events
- evaluation points
- coordinalization
- interval
- fan of evaluation points
- spatiotemporal locations
- an interpretation and a language

Let us explain which conditions are met and why.

**P-ord model** This is met by the BST model definition.

Flow of Time & Branching 197

**Chains** BST does have chains (or causal tracks).

**Alternatives** Obviously, alternatives are represented by histories.

**SLR** BST has a definition of SLR.

**Evaluation points** BST does not have evaluation points but they can be defined as a pair $e/h$ with $h$ being some history to which $e$ belongs.

**Coordinalization** BST lacks a coordinalization but it could be introduced based on worldlines. Those are dense and linearly ordered and we could create an order-preserving bijection between the worldline and a dense subset of some linearly ordered set.

**Interval** BST does not have intervals. Given a coordinalization, we could measure intervals based on worldlines.

**S-t locations** BST does not have s-t locations but we can introduce them a similar way as in BCont—a partition of $W$ with some specific properties.

**Fan of evaluation points** BST does not have this notion. However, there is no obstacle to introduce it to BST, especially if we have already introduced s-t locations.

**Interpretation and language** We can easily construct an interpretation. The language should be the same as in BCont and can be added to BST.

This sketch suggests that BST models could also incorporate the gFoT ideas we used in BCont. It is interesting to see that some ideas are missing in BST itself but they are present in MBS. This is, for example, the case of coordinalization (using Minkowski spacetime distance) or s-t locations (points from the Minkowski spacetime).

## 5 Discussion

The FoT approach taken in this article favours, as we have seen, the view that flow of time is merely an ordering and does not need anything more. However, there is a dichotomy present in our current work as we started out by having a flow of time that is merely an ordering but while evaluating for a given $e_C$ we do have an ontological difference between the future and the past. This can be seen on the settledness of the past but not of the future for a given point-event.

The other choice we made was to work with BCont rather than with BST or MBS. One could argue that MBS is more suitable to welcome a term from physics (as FoT). In spite of this we chose to use gFoT on BCont hoping it would be a challenging trial for gFoT and the use in BST/MBS would be simpler compared to the BCont.

## 6 Summary

We presented a way of how to interpret the idea of generalized flow of time by D. Dieks in Branching models. We have shown that it can be accommodated for the use in Branching Continuation models. First we verified their compliance with the original valuation from T. Placek and then we have shown how flow of time based valuation works in Branching Continuations. We demonstrated these properties on two examples, one analysing the question of Peircean future and the other examining the settledness of past events. Thereafter we sketched that this valuation would shift only a little in the Branching space-time model. Finally we closed with remarks on the nature of flow of time with regards to Branching Continuations.

## References

Belnap, N. (1992). Branching Space-Time. *Synthese, 92*, 385–434.
Belnap, N. (2003). *Branching space-time, postprint January, 2003*. Retrieved from http://philsci-archive.pitt.edu/1003/
Dieks, D. (1988). Special relativity and the flow of time. *Philosophy of Science, 55*, 456–460.
Hodkinson, I., & Reynolds, M. (2006). Temporal logic. In P. Blackburn, J. van Benthem, & F. Wolter (Eds.), *Handbook of modal logic* (Vol. 3, pp. 655–720). Amsterdam: Elsevier.
Placek, T. (2011). Possibilities Without Possible Worlds/Histories. *Journal of Philosophical Logic, 40*, 1–29.
Placek, T., & Belnap, N. (2011). Indeterminism is a modal notion: branching spacetimes and Earman's pruning. *Synthese, 187*, 1–29.
Placek, T., & Wroński, L. (2009). On Infinite EPR-like Correlations. *Synthese, 167*, 1–32.

Venema, Y. (2001). Temporal logic. In L. Goble (Ed.), *The Blackwell guide to philosophical logic* (pp. 203–223). Hoboken: Wiley-Blackwell.

Petr Švarný
Department of logic, Faculty of Arts, Charles University in Prague
Nám. Jana Palacha 2, 116 38 Praha 1, Czech Republic
e-mail: `svarnypetr@gmail.com`
URL: `https://sites.google.com/site/svarnypetr/`

# Cognitivist Probabilism

PAUL D. THORN

**Abstract:** In this article, I introduce the term "cognitivism" as a name for the thesis that degrees of belief are *equivalent* to full beliefs about truth-valued propositions. The thesis (of cognitivism) that degrees of belief are equivalent to full beliefs is equivocal, inasmuch as different sorts of equivalence may be postulated between degrees of belief and full beliefs. The simplest sort of equivalence (and the sort of equivalence that I will discuss here) identifies having a given degree of belief with having a full belief with a specific content. This sort of view was proposed in (Howson & Urbach, 1996). In addition to embracing a form of cognitivism about degrees of belief, Howson and Urbach argued for a brand of probabilism. I call a view, such as Howson and Urbach's, which combines probabilism with cognitivism about degrees of belief "cognitivist probabilism." In order to address some problems with Howson and Urbach's view, I propose a view that incorporates several of modifications of Howson and Urbach's version of cognitivist probabilism. The view that I finally propose upholds cognitivism about degrees of belief, but deviates from the letter of probabilism, in allowing that a rational agent's degrees of belief need not conform to the axioms of probability, in the case where the agent's cognitive resources are limited.

**Keywords:** bayesianism, probabilism, the Dutch book argument

## 1 Introduction

In their book *Scientific Reasoning: The Bayesian Approach* (1996), Howson and Urbach propose to treat an agent's beliefs about what betting quotients are fair for wagers as a measure of the agent's degrees of belief. More precisely, Howson and Urbach hold that if $r$ is the unique value such that an agent, $A$, believes that the betting quotient $r$ is fair for wagers on $\alpha$, then $A$'s degree of belief in $\alpha$ is $r$. To the preceding conditional, Howson and Urbach add the idealizing assumption that for each agent, $A$, and proposition, $\alpha$, there exists a unique $r$, such that $A$ believes that the betting quotient $r$ is fair for wagers on $\alpha$. Given their idealizing assumption, Howson and Urbach's proposed identification of degree of belief with a species of belief entails that an agent's degree of belief in $\alpha$ is $r$ if and only if $r$ is the unique value such that the agent believes that the betting quotient $r$ is fair for wagers on $\alpha$.[1]

With the proposed equivalence between degrees of belief and beliefs about betting quotients in the background, Howson and Urbach argue that having non-probabilistic degrees of belief (i.e., degrees of belief that do not conform to the axioms of probability) is equivalent to having an inconsistent set of beliefs about what betting quotients are fair. The idea is then that the prescription to have probabilistic degrees of belief (i.e., degrees of belief that conform to the axioms of probability) is a consequence of the prescription that agents should have consistent beliefs.

It is useful to think of Howson and Urbach's brand of probabilism as consisting of two theses. The first thesis is that degrees of belief are equivalent to beliefs about betting quotients (in the manner specified above). The second thesis is that it is inconsistent to regard an assignment of betting quotients as fair if those betting quotients do not conform to the axioms of probability. Howson and Urbach's first thesis reflects their subscription to cognitivism about degrees of belief, while the two theses together yield an argument for probabilism.

Howson and Urbach's argument for their second thesis shares features with traditional arguments for probabilism (cf. Ramsey, 1931, and de Finetti, 1937). Like traditional arguments for probabilism, Howson and Urbach's argument appeals to a consequence of the Dutch Book theorem. The relevant consequence states that if an assignment of betting quotients, $p$, fails to be a probability function, then there exists a possible set of wagers, $W$, in accordance with $p$, such that simultaneously accepting each element of $W$ is *assured* to result in a net loss. The Dutch Book theorem can be stated formally and subjected to rigorous mathematical proof. However, Howson and Urbach's argument for their second thesis relies on two further premises that cannot be subjected to mathematical proof. The two premises are: (1) if acceptance of a given set of wagers is assured to result in a net loss, then acceptance of the set is disadvantageous, and (2) acceptance of a set of wagers whose elements are each individually made at fair betting quotients is not disadvantageous. The second of these two premises has been the subject of criticism (cf. Maher, 1997). In order to avoid such criticisms, the view defended here does not rely on (2).

Howson and Urbach's first thesis depends on the assumption that, for

---

[1] In order for the entailment to hold it must also be assumed that if an agent's degree of belief in a proposition, $\alpha$, is $r$ and $r \neq s$, then it is not the case that the agent's degree of belief in $\alpha$ is $s$.

each agent, $A$, and proposition, $\alpha$, there is a unique betting quotient, $r$, that $A$ regards as fair for wagers on $\alpha$. Howson and Urbach acknowledge that attributing belief is such "advantage-equilibrating" betting quotients, in all cases, is a "strong" idealizing assumption (1996, p. 75). In order to address possible criticisms that may derive from this assumption, Howson and Urbach assert that the idealizing assumption can be relaxed. In the case where the assumption is relaxed (and interval-valued degrees of belief are defined in a manner analogous to point-valued degrees of belief), Howson and Urbach assert that consistent interval-valued degrees of belief conform to the usual principles that generalize the axioms of probability to interval-values.[2]

Within a cognitivist framework of the sort proposed by Howson and Urbach, I will propose a means to relaxing Howson and Urbach's idealizing assumption. In order to carry out the relaxation, I will appeal to the notions of *favorable* and *unfavorable* betting quotients, as a substitute for Howson and Urbach's notion of a *fair* betting quotient. By appeal to the notions of favorable and unfavorable betting quotients, I will endeavor to defend a view which evades several difficulties with Howson and Urbach view, while at the same time preserving the advantages of cognitivism about degrees of belief.[3]

## 2 Favorable and Unfavorable Betting Quotients

In this section, I provide a preliminary account of the notions of favorable and unfavorable betting quotients. For the sake of simplicity, the conditions are expressed relative to a propositional language, $L$, consisting of a countable set of propositional atoms, $p_1, p_2, p_3$, etc., and truth functional compounds of the atoms via the usual connectives: $\neg, \wedge, \vee, \rightarrow$ (material conditional), and $\leftrightarrow$ (material bi-conditional). Lower case Greek letters, $\alpha, \beta, \chi$, etc. will be used as meta-logical variables ranging over the elements of $L$. "$\models$" is used denote the (classical) logical consequence relation.

---

[2] In fact, Howson and Urbach's original argument for probabilism, which assumes point-valued degrees of belief, does not generalize so neatly to the case of interval-valued degrees of belief. Once one moves to the case of interval-valued degrees of belief, appeal to the Dutch Book theorem is insufficient to derive the conclusion that consistent interval-valued degrees of belief conform to the usual theorems that generalize the axioms of probability to interval-values (cf. Walley, 1991, p. 67).

[3] A significant disadvantage of Howson and Urbach view is its reliance on the package principle (cf. Schick, 1986; Hájek, 2008; Maher, 1997). Further problems with the view were outlined in (Vineberg, 2001).

In expressing the conditions for favorability and unfavorability, I will speak of wager *types* in addition to wager *tokens*. Among other things, this distinction will be used in expressing the fact that, in certain circumstances, agents are rationally required to have identical dispositions toward distinct tokens of the same wager type. Wager types are characterized as follows.

**Definition 1** *A wager type is a pair, $\langle \alpha, s \rangle$, where $\alpha$ is an element of $L$, and $s$ is real number.*

Throughout, wealth will be measured in a single currency, denoted "\$", and, for convenience, it is assumed that wagers have a standardized format:

Acceptance of an instance of a *standardized wager* of type $\langle \alpha, s \rangle$ *directly results* in an immediate +\$1-$s$ change in wealth, if $\alpha$ is true, and *directly results* in an immediate -\$$s$ change in wealth, if $\alpha$ is false. Acceptance of an instance of a *standardized wager* has no other *direct results*.[4]

For the purpose of expressing necessary and sufficient conditions for being favorable and unfavorable betting quotients, I will appeal to the notion of a *normalized agent*. Every normalized agent is assumed to possess a body of *primary evidence*, $E$, which is a set of sentences of $L$. $E$ is described as a normalized agent's 'primary evidence', in order to distinguish the contents of $E$ from other facts evident to normalized agents, as specified in condition 7 of the definition of *normalized agent* (see below). For a normalized agent, $A$, it is intended that $A$'s total evidence is comprised of $A$'s primary evidence in combination with the content specified by condition 7 of the following definition.

**Definition 2** *$A$ is a normalized agent with evidence, $E$, just in case:*

(1) *$A$'s utility function is a positive linear function of his/her final wealth state,*

(2) *$A$'s means of changing her wealth state is limited to the direct results of accepting standardized wagers,*

(3) *$A$ has (finite) wealth sufficient to accept all of the (finite number of) wagers that she is offered,*

(4) *$A$ has the opportunity to consider all of the wagers that she will be offered before deciding which wagers to accept,*

---

[4] Standardized wagers have a fixed stake of \$1. This limitation in the possible stakes of wagers is made for expository purposes. Wagers for greater stakes can be 'constructed' as multiple tokens of the same wager type.

*(5) A's has unlimited and unerring computational resources,*

*(6) A's primary evidence is E,*

*(7) A's total evidence consists of E, along with the knowledge that (1) through (6) hold of her, and the knowledge of what wagers she has been offered, and of all other propositions consequent to the full application of her computational resources, and*

*(8) A believes a proposition, p, iff that belief is supported by A's total evidence.*

The present definition is meant to characterize a sort of agent whose interests, abilities, and circumstances are sufficiently fixed, so that agents of the described sort are suitably homogeneous with respect to the prescriptions that apply to them vis-a-vis the acceptance and rejection of wagers.[5]

It is now possible to characterize relevant notions of favorable and unfavorable betting quotients.

**Definition 3** *$s$ is a favorable betting quotient for $\alpha$, given $E$, just in case for all possible $A$ and $w$, if $A$ is a normalized agent with evidence, $E$, and $w$ is a wager of type $\langle \alpha, s \rangle$, and $A$ is offered $w$, then $A$ is rationally required to accept $w$.*[6]

**Definition 4** *$s$ is an unfavorable betting quotient for $\alpha$, given $E$, just in case for all possible $A$ and $w$, if $A$ is a normalized agent with evidence, $E$, and $w$ is a wager of type $\langle \alpha, s \rangle$, and $A$ is offered $w$, then $A$ is rationally required to reject $w$.*

Although the conditions for being favorable and unfavorable betting quotients are characterized as definitions, it is preferable to think of the 'definitions' as specifying necessary and sufficient conditions that hold in all possible worlds. Indeed, it is clear, for both conditions, that the proposed analysans outstrips the content of the ordinary intuitive conception of what it is for a betting quotient to be favorable or unfavorable.

---

[5] A similar kind of definition is proposed for a similar purpose in (Christensen, 1996) (cf. Christensen, 2001, 2004).

[6] The sort of rationality appealed to in the definition is inclusive of practical and epistemic rationality. In other words, any agent who fails to abide by the rationality requirement invoked by the condition is guilty of either practical or epistemic irrationality.

## 3 Postulates concerning Favorability and Unfavorability

The goal of the present section is to propose and defend a series of postulates that partially characterize the nature of favorable and unfavorable betting quotients. In the following section, the principles will be used in showing that betting quotients that are neither favorable nor unfavorable conform to the axioms of probability. I begin by listing seven postulates that are conceptually simple, and (hopefully) uncontroversial, given the conditions for favorability and unfavorability that I stipulated in the previous section.

**Postulate 5** (Complete Possibilities) *For every set, $E$, of sentences of $L$, and for every ordered set of wager types, $\langle W_1, W_2, \ldots, W_n \rangle$, there exists a possible normalized agent with evidence, $E$, such that the set of wagers offered to $A$ is $\langle w_1, w_2, \ldots, w_n \rangle$, where for all $i$, $w_i$ is an instance of $W_i$.*[7]

**Postulate 6** (Consistency of Prescriptions) *$s$ is a favorable betting quotient for $\alpha$, given $E$, only if $s$ is not an unfavorable betting quotient for $\alpha$, given $E$.*

**Postulate 7** (Opposition) *For all $r$, $\alpha$, and $E$, $r$ is a favorable betting quotient for $\alpha$, given $E$, just in case $1 - r$ is an unfavorable betting quotient for $\neg\alpha$, given $E$.*

Note that a wager of type $\langle \neg\alpha, 1 - r \rangle$ simply is the 'other side' of a wager of type $\langle \alpha, r \rangle$. So postulate 7 says that one side of a wager is favorable just in case the other is unfavorable (assuming fixed evidence $E$).

**Postulate 8** (Sure Gain) *If acceptance of a standardized wager of type $\langle \alpha, s \rangle$ directly results in an increase in wealth in all possible situations consistent with $E$, then $s$ in a favorable betting quotient for $\alpha$, given $E$.*

**Postulate 9** (Sure Loss) *If acceptance of a standardized wager of type $\langle \alpha, s \rangle$ directly results in a decrease in wealth in all possible situations consistent with $E$, then $s$ is an unfavorable betting quotient for $\alpha$, given $E$.*

**Postulate 10** (Universality) *For all possible $A, E, w$, and $W$, if $A$ is a normalized agent with evidence $E$, and $A$ is offered $w$ (a wager of type $W$), then if $A$ is rationally required to accept/reject $w$, then for all possible $A'$*

---

[7] It is not assumed that $i \neq j$ implies $W_i \neq W_j$, though it is, of course, assumed that $i \neq j$ implies $w_i \neq w_j$.

and $w'$, if $A'$ is a normalized agent with evidence $E$, and $A'$ is offered $w'$ (a wager of type $W$), then $A'$ is rationally required to accept/reject $w'$.[8]

In essence, postulate 10 says that normalized agents with identical evidence are indistinguishable with respect to the wagers they are required to accept (and reject). The postulate is justified, since the characterization of normalized agents sufficiently fixes the nature and circumstances of such agents, so they cannot differ in a way that is relevant to which types of wagers they are required to accept (and reject). Note especially that condition 8 of the definition of normalized agent assures uniformity in what normalized agents with identical evidence believe. In effect, normalized agents are incredulous, and only believe propositions that are supported by their evidence.

**Postulate 11** (Extensionality)    *For all $E, W_1, W_2,$ and $W_3$, if, in all possible situations consistent with $E$, the payoff for an instance of a wager of type $W_1$ is identical to the sum of the payoffs for an instance of a wager of type $W_2$ and an instance of a wager of type $W_3$, then for all possible $A, w_1, w_2,$ and $w_3$, if $A$ is a normalized agent with evidence $E$, and $A$ is offered $w_1$ (a wager of type $W_1$), $w_2$ (a wager of type $W_2$), and $w_3$ (a wager of type $W_3$), then: if $A$ is rationally required to accept/reject $w_2$ and $A$ rationally required to accept/reject $w_3$, then $A$ is rationally required to accept/reject $w_1$.*

Postulate 11 is a watered-down cousin of the package principle (cf. Hájek, 2008). While the package principle is dubitable, Postulate 11 is sufficiently limited as to be unassailable. The operative content of the principle tells us that if a pair of wagers $w_1$ and $w_2$ are jointly equivalent to a wager $w_3$, then if a normalized agent is required to accept (reject) both elements of the pair, then the agent is required to accept (reject) the single wager that is equivalent to the pair.

## 4    The Axioms of Probability

The point of the present section is to show that, relative to any body of evidence, $E$, the set of betting quotients that are neither favorable nor unfavorable conform the axioms of probability. The following definitions will

---

[8]The condition expressed by the postulate is meant to hold in the case where "accept" or "reject" is substituted for "accept/reject".

be used in expressing the theorems which correspond to the axioms of probability.

**Definition 12**  $P_E(\alpha) = \{s | s \text{ is neither a favorable nor unfavorable betting quotient for } \alpha, \text{ given } E \}$.

**Definition 13**  $P\lfloor_E(\alpha) = supremum\{s | s \text{ is a favorable betting quotient for } \alpha, \text{ given } E \}$.

**Definition 14**  $P\lceil_E(\alpha) = infimum\{s | s \text{ is a unfavorable betting quotient for } \alpha, \text{ given } E \}$.

With no further assumptions (save Postulates 5 through 11), we may prove the following theorems that correspond to the standard Kolmogorov axiomatization of probability (without countable additvity).

**Theorem 15**  $\forall \alpha, E : P\lfloor_E(\alpha) \geq 0$.

**Theorem 16**  $\forall \alpha, E : \models \alpha \Rightarrow P_E(\alpha) = \{1\}$.

**Theorem 17**  $\forall \alpha, \beta, E : \{\alpha\} \models \neg \beta \Rightarrow P\lfloor_E(\alpha) + P\lfloor_E(\beta) \leq P\lfloor_E(\alpha \vee \beta) \leq P\lceil_E(\alpha \vee \beta) \leq P\lceil_E(\alpha) + P\lceil_E(\beta)$.

**Theorem 18**  $\forall \alpha, E : P\lfloor_E(\neg \alpha) = 1 - P\lceil_E(\alpha)$.

One thing that we cannot prove, for arbitrary $E$ and $\alpha$, is that $P_E(\alpha) \neq \emptyset$. One way to incorporate this assumption is to introduce an additional postulate.

**Postulate 19** (Existence)  $\forall \alpha, E : P_E(\alpha) \neq \emptyset$.

Postulate 19 corresponds to an assumption that is far weaker than the sort of one that is generally made in discussions of degree of belief (namely, that rational degrees of belief are representable by either point- or interval-values). Yet while arguments can be given on behalf of the Postulate 19, I am willing to concede that it is dubitable. For one, it not obviously incoherent that there be a case where all of the betting quotients for a given proposition are either favorable or unfavorable. One simply imagines that all betting quotients up to some value, $s$, are favorable, and that all betting quotients greater than $s$ are unfavorable. In any case, I will not fuss over Postulate 19, since the loss in the case where the postulate is rejected is, literally, infinitesimal. Nevertheless, assuming Postulate 19 (in addition to Postulates 5 through 11), one can prove the following:

**Theorem 20**  $\forall E : \exists p : p$ *is a probability function and* $\forall \alpha : p(\alpha) \in P_E(\alpha)$.

The connection between Theorems 15, 16, 17, and 18, and the axioms of probability become clearest, if we make the additional supposition, for all $E$ and $\alpha$, that there exists an $r$, such that $P_E(\alpha) = \{r\}$. In that case, we can recover Kolmogorov's axioms for functions $P'_E$, where we suppose $P'_E(\alpha) = r$ just in case $P_E(\alpha) = \{r\}$.[9]

## 5 Imperfect Agents

So far, we have seen, given a particular understanding of what it is for a betting quotient to be favorable or unfavorable, that betting quotients that are neither favorable nor unfavorable conform to the axioms of probability. However, given the proposed understanding of favorability and unfavorability, there appear to be cases where the rational degree of belief in a proposition diverges from corresponding rational beliefs about what betting quotients are neither favorable nor unfavorable for wagers on the proposition. For example, in the case of a mathematical statement, like Goldbach's Conjecture, it appears that a rational agent (in circumstances such as ours) will believe that there is one betting quotient for the Conjecture that is neither favorable nor unfavorable, and that that betting quotient is either *zero* or *one* (cf. Vineberg, 2001). At the same time, the agent may have a high degree of belief, $r$, in the Conjecture, where $r < 1$.

In order to bridge the connection between rational degrees of belief and rational beliefs about betting quotients for agent's with limited cognitive resources, I will now generalize the notions of favorability and unfavorability that were proposed in section 2. The first step toward expressing these generalizations is to enrich the language that is used to describe a normalized agent's body of primary evidence. To keep things simple, the revised definition of *normalized agent* will be expressed relative to a language $L'$, which is formed by enriching the propositional language, $L$, by addition of the set of expressions of the form: $S \models \alpha$ (where $\alpha$ is a sentence of $L$, and where $S$ is a set of sentences of $L$). The revised definition of *normalized agent*, then assumes that every normalized agent possesses a body of primary evidence,

---

[9] In the interests of space, I will not present a generalization of the notion of favorable and unfavorable betting quotients for application to conditional wagers. Such a generalization is possible, and given some reasonable postulates about conditional wagers, it is trivial to prove a theorem that corresponds to the standard definition of conditional probability.

$E$, which is a set of sentences of $L'$. We must also modify the definition of *normalized agent* (originally provided in section 2), so that it is no longer assumed that normalized agents have unlimited computational and ratiocinative resources.

**Definition 21**   *A is a normalized\* agent with evidence, E, just in case:*

 (1) *A's utility function is a positive linear function of his/her final wealth state,*

 (2) *A's means of changing his/her wealth state is limited to the direct effects of accepting standardized wagers,*

 (3) *A has wealth sufficient to accept all of the wagers that he/she is offered,*

 (4) *A has the opportunity to consider all of the wagers that he/she is offered before deciding which wagers to accept,*

 (5) *A's primary evidence is E,*

 (6) *A's accessible evidence consists of E, along with the apprehension of what wagers he/she has been offered, and the apprehension that (1) through (5) hold, and*

 (7) *A believes a proposition, $\alpha$, just in case $\alpha$ is included in A's accessible evidence.*

It is not assumed, in general, that $E$ is closed under logical consequences. Rather $E$ will be understood to represent the set of propositions that are 'accessible' to $A$. The precise content of the present notion of accessibility is left relatively open. The general idea is that some contents implicit in an agent's evidence are accessible, while other implicit contents (conclusions that can only be reached via long derivations from accessible contents, for example) are inaccessible.

Corresponding to the definition of *normalized\** agent, the conditions for $s$ being a *favorable\** or *unfavorable\** betting quotient are identical to the conditions for $s$ being a favorable or unfavorable betting quotient, save that the conditions now apply to *normalized\** rather than *normalized* agents.

The postulates that were formerly used in characterizing that nature of favorable and unfavorable betting quotients must be revised before they are fit to characterize the nature of favorable* and unfavorable* betting quotients. The contents of Postulates 5, 6, 7, and 10 are still reasonable, when expressed in terms of *favorable** and *unfavorable** betting quotients, and remain unchanged, save that they are now expressed in terms of *favorable** and *unfavorable** wagers, and the language $L'$. The remaining postulates are modified, as follows.

**Postulate 22** (Sure Gain*)  *If $s < 0$ or ($\emptyset \models \alpha \in E$ and $s < 1$), then $s$ is a favorable* betting quotient for $\alpha$, given $E$.*

**Postulate 23** (Sure Loss*)  *If $s > 1$ or ($\emptyset \models \neg\alpha \in E$ and $s > 0$), then $s$ is an unfavorable* betting quotient for $\alpha$, given $E$.*

**Postulate 24** (Extensionality*)  *For all $E, \alpha, \beta, \chi, s_\alpha, s_\beta$, and $s_\chi$, if $\{\alpha\} \models \neg\beta \in E$ and $\alpha \vee \beta = \chi$ and $s_\alpha + s_\beta = s_\chi$, then for all possible $A, w_\alpha, w_\beta$, and $w_\chi$, if $A$ is a normalized agent with evidence $E$, and $A$ is offered $w_\alpha$ (a wager of type $\langle \alpha, s_\alpha \rangle$), $w_\beta$ (a wager of type $\langle \beta, s_\beta \rangle$), and $w_\chi$ (a wager of type $\langle \chi, s_\chi \rangle$), then: if $A$ is rationally required to accept/reject $w_\alpha$ and $A$ rationally required to accept/reject $w_\beta$, then $A$ is rationally required to accept/reject $w_\chi$.*

Next, the functions $P\lfloor_E(\alpha), P\lceil_E(\alpha)$, and $P_E(\alpha)$ must be amended to form the functions $P^*\lfloor_E(\alpha), P^*\lceil_E(\alpha)$, and $P^*_E(\alpha)$, by substituting instances of "*favorable**" and "*unfavorable**" for all instances of "*favorable*" and "*unfavorable*". In that case, the following theorems hold (given Postulates 22, 23, and 24, along with suitably modified versions of Postulates 5, 6, 7, and 10).

**Theorem 25**  $\forall \alpha, E : P^*\lceil_E(\alpha) \geq 0$.

**Theorem 26**  $\forall \alpha, E : \emptyset \models \alpha \in E \Rightarrow P^*_E(\alpha) = \{1\}$.

**Theorem 27**  $\forall \alpha, \beta, E : \{\alpha\} \models \neg\beta \in E \Rightarrow P^*\lfloor_E(\alpha) + P^*\lfloor_E(\beta) \leq P^*\lfloor_E(\alpha \vee \beta) \leq P^*\lceil_E(\alpha \vee \beta) \leq P^*\lceil_E(\alpha) + P^*\lceil_E(\beta)$.

**Theorem 28**  $\forall \alpha, E : P^*\lfloor_E(\neg\alpha) = 1 - P^*\lceil_E(\alpha)$.

Theorems 25, 26, 27, and 28 correspond to Kolmogorov's axioms. The four theorems also describe some basic constraints on which betting quotients can be neither favorable* nor unfavorable*, given a body of evidence, $E$. These constraints have the status of conceptual truths, so that a set of beliefs about betting quotients that are inconsistent with the theorems is thereby inconsistent.

## 6 Conclusion

Theorems 25, 26, 27, and 28 embody a plausible view of the constraints that the axioms of probability place on rational belief. Some consequences of the theorems are as follows:

1. For all propositions, $\alpha$, it is *inconsistent* to have a degree of belief $R$ in $\alpha$, if there exists an $r$ in $R$ and $r < 0$.

2. For all propositions, $\alpha$, if it is *evident* that $\alpha$ is a logical truth, then if $R \neq \{1\}$, then it is *inconsistent* to have a degree of belief $R$ in $\alpha$.

3. For all propositions, $\alpha$ and $\beta$, if it is evident that $\alpha$ and $\beta$ are logically inconsistent, then, for all sets $R_\alpha, R_\beta$, and $R_{\alpha \vee \beta}$, if $inf(R_\alpha) + inf(R_\beta) > inf(R_{\alpha \vee \beta})$ or $sup(R_\alpha) + sup(R_\beta) < sup(R_{\alpha \vee \beta})$, then it is *inconsistent* to simultaneously have a degree of belief $R_\alpha$ in $\alpha$, degree of belief $R_\beta$ in $\beta$, and degree of belief $R_{\alpha \vee \beta}$ in $\alpha \vee \beta$.

I have defended cognitivism about degrees of belief, and view which deviates a little from probabilism. Call the view defended "cognitivist probabilism*." Cognitivist probabilism* is immune to a number of objections to Howson and Urbach's brand of probabilism. At the same time, the view maintains the principal advantages of cognitivist probabilism, with adjustments made to account for the proposed view's deviation from the letter of probabilism. First, the demand that one's degrees of belief conform to the axioms of probability (modulo one's ability to detect certain inferential relations between propositions) is a consequence of the demand that one have consistent beliefs. Second, we can understand the prescription that degrees of belief conform to the axioms of probability (modulo one's ability to detect certain inferential relations between propositions) independently of specifying the prescriptive connection between degrees of belief and action. Third, given the postulated equivalence between degrees of belief and corresponding full beliefs, degrees of belief are brought within the fold of epistemology and logic.[10]

---

[10] Work on this paper was supported by the DFG financed EuroCores LogiCCC project *The Logic of Causal and Probabilistic Reasoning in Uncertain Environments*, and the DFG project *The Role of Meta-Induction in Human Reasoning* (SPP 1516). For valuable comments I am indebted to Ludwig Fahrbach, Gerhard Schurz, Hannes Leitgeb, Peter Milne, and audiences at LOGICA 2012, the Heinrich-Heine-Universität Düsseldorf, and 7th Annual Formal Epistemology Workshop in Konstanz.

# References

Christensen, D. (1996). Dutch-book arguments depragmatized: Epistemic consistency for partial believers. *The Journal of Philosophy, 93*, 450–479.

Christensen, D. (2001). Preference based arguments for probabilism. *Philosophy of Science, 68*, 356–376.

Christensen, D. (2004). *Putting logic in its place: Formal constraints on rational belief.* Oxford: Oxford University Press.

de Finetti, B. (1937). La prévision: Ses lois logiques, ses sources subjectives. *Annales de l'Institut Henri Poincaré, 7*, 168. (translated as de Finetti, 1980)

de Finetti, B. (1980). Foresight. Its logical laws, its subjective sources. In J. H. E. Kyburg & H. E. Smokler (Eds.), *Studies in subjective probability.* Robert E. Krieger Publishing Company.

Hájek, A. (2008). Arguments for—or against—probabilism? *Brit. J. Phil. Sci., 59*, 793–819.

Howson, C., & Urbach, P. (1996). *Scientific reasoning: The bayesian approach* (2nd ed.). Chicago: Open Court.

Maher, P. (1997). Depragmatized dutch book arguments. *Philosophy of Science, 64*, 291–305.

Ramsey, F. (1931). *Truth and probability, in foundations of mathematics and other essays.* Routledge & P. Kegan.

Schick, F. (1986). Dutch bookies and money pumps. *Journal of Philosophy, 83*, 112–119.

Vineberg, S. (2001). The notion of consistency for partial belief. *Philosophical Studies, 102*, 281–296.

Walley, P. (1991). *Statistical reasoning with imprecise probabilities.* Chapman and Hall.

Paul D. Thorn
Institut für Philosophie
Heinrich-Heine-Universität
Universitätsstr. 1
D-40204, Düsseldorf, Germany
e-mail: thorn@phil.hhu.de

# An Analogy in Dummett's Views on Truth- and Proof-conditional Meaning Theories

### LUCA TRANCHINI[1]

**Abstract:** I argue that both in the account of meaning based on truth-conditions and in the one based on proof-conditions the specification of the central semantic notion relies on an ancillary one. In one case, understanding the truth-conditions of quantified sentences relies on knowing the conditions at which a predicate is true of an arbitrary object; in the other case, understanding the canonical proof-conditions of sentences relies on the ability of recognizing hypothetical deducibility relations among them. Dummett claims that the ancillary notions are irrelevant for characterizing speakers' understanding. I will argue, on the contrary, that the ancillary notions play a more prominent role than Dummett is willing to accord them.

**Keywords:** Frege, implication, predicate, satisfaction, canonical proof

## 1 Frege, Tarski and Dummett

In a propositional language, given a specification of the truth-value of atomic sentences, the truth-value of logically complex sentences is defined in terms of that of their sub-sentences. As Tarski's analysis shows, in a first-order setting truth cannot be inductively defined.[2] What can be inductively defined is the notion of satisfaction of a (possibly open) formula by a sequence of objects (or say, by an assignments of objects for the variables of the language). Truth of sentences (closed formulas) is achieved by means of a 'blanket definition': a sentence is true if and only if (iff) it is satisfied by all sequences (assignments).

However, it is usually stressed that truth—and not satisfaction—is the central semantic notion. For instance Dummett, in comparing Tarski's and

---

[1]This work is supported by the German Research Agency (DFG grant Tr1112/1) as part of the project 'Logical consequence. Proof-theoretic and epistemological perspectives'.

[2]Unless one is willing to adopt a substitutional interpretation of quantification. In the present context, this will not be considered as an option.

Frege's views, argues that open formulas correspond in the last instance to ($n$-ary) predicates. And the category of sentences and singular terms is prior to the one of predicates since, apart from the primitive ones belonging to the vocabulary of the language, predicates are in general obtained from complete sentences by removing one or more occurrences of one or more singular terms.

## 1.1 Predicates

For example, the sentence '$0 + 7 = 7$' is obtained by chaining together '0' and '7' by means of the functional expression '+', and the two terms '$0+7$' and '7' with '='. The meaning of the sentence is obtained from that of its components and, in the end, from the primitive expressions which constitute it. The sentence expresses the identity of two numerical terms ('$0 + 7$' and '7') and it is true because they refer to the same object.

Alternative ways of analysing the sentence arise by tracing what we will call 'complex predicative patterns' within the sentence. An example is the predicate '$0 + \xi = \xi$', which is true of an object $o$ iff the result of adding $o$ to 0 is identical with $o$. To recognize this complex predicate within the sentence is to analyse the sentence as stating that the number 7 enjoys the property expressed by the predicate.

Dummett refers to these as "two different kinds of analysis of a sentence into its constituents" (Dummett, 1981, pp. 64–65): The sentence is either viewed as composed out of the language primitive expressions, or as obtained by filling the slot of the complex predicate respectively.

According to him, only the first one is

> "an analysis which relates to sense. The other kind of analysis is needed in order to determine the validity of inferences in which the sentence may be involved, and it is unnecessary, for someone to understand the sentence, that he be aware of the possibility of an analysis of this kind." (*ibid.*, 65)

Dummett claims that the ability to recognise all the complex predicative patterns contained in a sentence is not necessary for understanding it. However, he also stresses that this ability is necessary for the understanding of more complex sentences.

According to Frege, a quantified sentence is obtained by applying a quantifier to a predicate (in general, not necessarily a primitive predicate, but a possibly complex one obtained by the operation of removing one or

# Truth-conditional and Proof-conditional Meaning Theories

more occurrences of a term from a complete sentence). The truth of the quantified sentence depends on which objects the predicate which constitutes the sentence is true of. For instance, a universally quantified sentence is true iff the predicate is true of every object.

To dwell on the previous example, in order to form the quantified sentence '$\forall x(0 + x = x)$' one needs first to recognize the complex predicate '$0 + \xi = \xi$' in its instances; to understand its truth-conditions one needs first to grasp what it is for the complex predicate '$0 + \xi = \xi$' to be true of an arbitrary object.

## 1.2 Frege's Assumption and Tarski's Truth Definition

In the following I will express the fact that a speaker knows what it is for a predicate to be true of an arbitrary object by saying that the speaker 'masters the predicate', and I will call such knowledge 'the mastery of the predicate'.

Hence, for Dummett, whereas the mastery of a complex predicate is needed to understand a sentence obtained by applying a quantifier to it, such a mastery is not needed to understand the instances of the quantified sentence from which the predicate has been extracted.[3] Recall his point: "It is unnecessary, for someone to understand the [instances], that he be aware of the possibility of [analysing]" them as constituted by the complex predicate.

By contraposition, it is possible that someone understands the instances *without* mastering the complex predicate.

However, in order to explain speakers' competence of quantified sentences, we must ascribe them the mastery of the predicate. If, as Dummett claims, the mastery of a predicate is not an ingredient of the understanding of the sentences from which the predicate is extracted, it is reasonable to ask where does this mastery come from.

For Frege, such a question does not arise at all. According to Dummett, what underlies Frege's account is the

> "assumption that, whenever we understand the truth-conditions for any sentence containing (one or more occurrences of) a proper name, we likewise understand what it is for any arbitrary object to satisfy the predicate which results from removing (those occurrences of) that proper name from the sentence." (*ibid.*, p. 17)

If understanding a sentence is equated with understanding its truth-conditions, Frege's assumption contradicts Dummett. The assumption amounts

---

[3] Unless, of course, the complex predicate does 'coincide' with one of the simple predicates composing the sentence.

to the following implication: If one understands the truth-conditions of a sentence then she also grasps what it is for all complex predicate (contained in the sentence) to be true of an arbitrary object. That is, Dummett ascribes Frege the view that the mastery of all complex predicates contained in a sentence *is* indeed a necessary condition for the understanding of the truth-conditions of a sentence, that is for understanding the sentence itself.[4]

To avoid Frege's assumption, the mastery of predicates may be viewed as a primitive ability which constitutes an extra-ingredient of speakers' competence. In other words, understanding a sentence can be equated with understanding its truth-conditions *and* mastering all the complex predicates it contains. Clearly, also this contradicts Dummett's claim that mastery of all the complex predicates traceable in a sentence is unnecessary for its understanding.

Furthermore, once this extra-ingredient has been introduced, it actually supersedes the other. As a matter of fact, not only the truth-conditions of quantified sentences, but of any sentence can be recovered from the notion of a predicate's being true of an arbitrary object. Hence, speaker's competence can be characterized as consisting *only* in the mastery of complex predicates.

To appreciate this, one should remember that complex predicates are nothing but open formulas. Tarski's definition of the truth of sentences in terms of the notion of satisfaction (of open formulas by sequences, or assignments, of objects) shows how truth-conditions of *all* sentences can be given in terms of the notion of a predicate's being true of an arbitrary object.

### 1.3 Sentences' Dual Role

What is then of Dummett's claim that the mastery of all complex predicates contained in a sentence is unnecessary for understanding it? Even if one were to assume that understanding a sentence merely means understanding its truth-conditions, as soon as one looks at complex sentences, in particular at quantified sentences, one would be forced to either of the following:

(i) assume that understanding the truth-conditions of a sentence implies the mastery of the complex predicates it contains;

(ii) take also the mastery of complex predicates as an ingredient of sentences' understanding.

---

[4]Throughout the paper, I take the fact that $B$ is a necessary condition for $A$ (and that $A$ is sufficient for $B$) to be expressed by the implication 'if $A$ then $B$'. Although not the only possibility, it is certainly the standard one (Brennan, 2011).

Clearly, both (i) and (ii) contrast with Dummett's claim. The idea of a "contrast between understanding a sentence as used on its own [...], and understanding it as capable of being a constituent in a more complex sentence." (*ibid.*, 449) is a pervasive theme of Dummett's interpretation of Frege. Although a fully-fledge evaluation of this idea is way beyond the scope of the present paper, I hope the reader is now convinced that Dummett perceives a tension in the truth-conditional account of meaning between two aspects of the semantic role of a sentence: its truth-conditions and the contributions to the truth-conditions of quantified sentences constituted by the complex predicates it contains.

In the remaining part of the paper, I argue that an analogous contrast is found also in the proof-theoretic account of meaning which Dummett proposes as an alternative to the one based on the notion of truth.

## 2 From Truth to Proofs

### 2.1 Kinds of Analysis and Ways of Establishing Sentences

I take a step back to reconsider one aspect of Dummett's remark on the different kinds of analysis which has not been touched yet.

For Dummett, the kind of analysis yielded by the recognition of complex predicates within sentences is "needed in order to determine the validity of inferences in which the sentence may be involved."

To clarify this point it is useful to reconsider the example discussed above. The analysis of '$0 + 7 = 7$' as resulting by filling with '7' the slot in the predicate '$0 + \xi = \xi$' is what enables the recognition of the validity of the inference from '$\forall x(0 + x = x)$' to '$0 + 7 = 7$'. If we did not recognize the complex predicate '$0 + \xi = \xi$' within '$0 + 7 = 7$', we would not be able to recognize the inference as an application of universal instantiation ($\forall$-elim).

Equivalently, whenever we recognize a sentence as containing a complex predicate, we thereby envisage the possibility of establishing it by a particular means (in our example, by universal instantiation from '$\forall x(0 + x = x)$', the sentence saying that zero is a (left-)identity with respect to addition).

Forgetting for a moment the entanglement of truth-conditions and the mastery of the totality of predicates contained in a sentence stressed in the previous section, and assuming Dummett's stance, according to which mastery of predicates is unnecessary for understanding it, the following considerations arise.

The way of establishing '$0 + 7 = 7$' just described requires the (at least implicit) mastery of the concept of identity element for addition (not to speak of the principle of mathematical induction which must be used to establish the universally quantified claim). And, intuitively at least, it would be tempting to say that mastery of the concept of identity element for addition is not required for understanding a sentence such as '$0 + 7 = 7$' (although it is even intuitively required for understanding the quantified sentence '$\forall x(0 + x = x)$').[5]

Now, the truth-conditions of a sentence may also be viewed as encoding a particular way of establishing its truth-value: Namely, by evaluating the numerical expressions '$7 + 0$' and comparing the result with 7. Given the *prima facie* plausible identification of sense with truth-conditions, to establish the sentence in accordance with its sense thus seems to require only the mastery of concepts one must already possess in order to understand the sentence.

I will refer to this as the *direct* or *canonical* way of establishing the truth-value of the sentence. The alternative ways of doing this suggested by the predicates traceable within the sentence will be referred to as indirect or non-canonical.

## 2.2 Two Kinds of Proofs

I propose to identify the means of establishing the truth-value of a sentence with a proof. The different kinds of analysis of a sentence yield a distinction between direct and indirect means of establishing a sentence, and hence a distinction between two kinds of proofs: Direct or canonical proofs and indirect or non-canonical proofs.

Dummett claims that awareness of only one kind of analysis is necessary for the understanding of a sentence, namely the analysis in terms of its component expressions. Given the correspondence between different kinds of analysis and different kinds of proof, a parallel claim can be made regarding the relationship of the two kinds of proofs to understanding. For Dummett:

> "The possibility of establishing the statement directly must be envisaged by anyone who grasps the meaning of the statement [...T]he possibility of establishing it indirectly need not be." (Dummett, 1973, pp. 312–313)

---

[5] Given that concepts are for Frege the semantic correlates of predicates, this is nothing but rephrasing Dummett's above considerations.

As the analyses of a sentence yielded by the complex predicates it contains are unnecessary for understanding it, so are the indirect proofs of it yielded by such a kind of analysis. Only direct or canonical proofs are connected with the meaning of sentences.

Dummett and others propose an alternative to the truth-condi-tional account of meaning in which the notion of proof replaces the one of truth as the central concept of the explanation. The different role played by the two kinds of proof with respect to understanding is one of the basic tenets of this alternative account.

This is usually expressed, for example by Prawitz (2006), by saying that the account of meaning should not be given in terms of the conditions of provability *tout court*, but rather in terms of the conditions of canonical provability: "The meaning of a sentence is given by what counts as a canonical proof of the sentence." (p. 515)

As we saw, Dummett perceives a tension between the semantic role of a sentence when viewed on its own and when viewed as a component of more logically complex ones. In particular, in order to account for the understanding of quantified sentences, the claim that the mastery of complex predicates is unnecessary for understanding must go.

In the next section, I argue that an analogous tension arises when the account of meaning is recast in terms of proof-conditions. In particular, the claim that only knowledge of the canonical proof-conditions is necessary for understanding will be hard to maintain.

## 3 Proof-theoretic Semantics

### 3.1 Natural Deduction

Let's assume that proofs are faithfully represented in the format of natural deduction. In the natural deduction system for intuitionistic logic NI, every logical constants † has associated a set of introduction rules (†-intro) and a set of elimination rules (†-elim).[6]

In a schematic presentation of a †-intro rule, the only occurrence of a logical constant is an occurrence of † as the main connective of the conclusion.

Introduction rules of the most simple kind can be specified by making reference to the immediate premises of the rules and to its conclusion only.

---

[6] I take negation as defined through implication and absurdity, and the *ex falso* rule as an elimination rule.

For instance, the conjunction introduction rule ($\wedge$-intro) looks as follows:

$$\frac{A \quad B}{A \wedge B} \wedge\text{-intro}$$

Not all introductions are of such a simple kind. The introduction rule for implication ($\rightarrow$-intro) makes reference not only to the immediate premise of the rule, but also to the whole deduction having the premise as conclusion:

$$\frac{\begin{array}{c}[A]\\ \mathscr{D}\\ B\end{array}}{A \rightarrow B} \rightarrow\text{-intro}$$

The reason is that the introduction rule, as one says, 'discharges' an assumption. That is, by applying the rule, one obtains a new conclusion ($A \rightarrow B$) which does not depend on the assumption $A$ upon which the immediate premise of the rule $B$ depended.

If in a deduction all assumptions are discharged, the conclusion is established categorically. With 'proof' I will thereby mean deductions in which all assumptions are discharged.

The set of elimination rules is obtained by 'inverting' the set of introduction rules according to the so-called inversion principle:

> "The conclusion obtained by an elimination does not state anything more than what must have already been obtained if the major premiss of the elimination was inferred by an introduction." (Prawitz, 1971, p. 246)

In the case of conjunction, we have two elimination rules ($\wedge$-elim$_1$ and $\wedge$-elim$_2$), which respectively permit to infer either of the conjuncts from the conjunction:

$$\frac{A \wedge B}{A} \wedge\text{-elim}_1 \qquad \frac{A \wedge B}{B} \wedge\text{-elim}_2$$

The inversion principle suggests that the patterns resulting by applying an elimination rule immediately after (one of) the corresponding introduction rule(s) constitute a roundabout in the proof. In the case of conjunction, one of the patterns in question is the following:

$$\frac{\dfrac{\begin{array}{cc}\mathscr{D}_1 & \mathscr{D}_2\\ A & B\end{array}}{A \wedge B} \wedge\text{-intro}}{A} \wedge\text{-elim}_1 \quad \stackrel{\wedge-\text{Red}_1}{\Longrightarrow} \quad \begin{array}{c}\mathscr{D}_1\\ A\end{array}$$

Suppose $\mathscr{D}_1$ and $\mathscr{D}_2$ are proofs of $A$ and $B$ respectively. Then the conclusion of $\wedge$-elim$_1$ was already established by $\mathscr{D}_1$.

Normalization theorems show that every pattern of this kind can be eliminated from deductions. This is done by defining for each connective so-called reduction procedures which show how to get rid of these redundancies (one of the reductions for conjunction, $\wedge$-Red$_1$, is shown above).

## 3.2 The Primacy of Introductions

The inversion principle is usually taken as expressing a certain primacy of the introduction over the eliminations. This can be formulated by saying that the introduction rules specify the meaning of logical constants while elimination rules are just consequences of this specification.[7]

Hence, given proofs for atomic sentences, a proof of a logically complex sentence which proceeds only by introduction rules is one which is shaped step by step in accordance with the meaning of the sentence.

As Dummett observes,

> "so far as a logically complex statement is concerned, the introduction rules governing the logical constants occurring in the statement display the most direct means of establishing the statement, step by step in accordance with its logical structure." (Dummett, 1975, p. 12)

If we restrict the attention to a language in which the meaning of logical constants is given by introduction rules of the simplest kind, it is natural to identify canonical proofs with deductions using only introduction rules (to be applied to proofs of the atomic sentences). And the canonical proofs of complex sentences can be specified in terms of canonical proofs of their components.[8]

This picture thus adheres to Dummett's claim that grasp of meaning only requires to envisage "the possibilty of establishing a statement directly". The canonical proofs of a sentence represent the most direct means of establishing it. And it is in their terms that the meaning of a sentence is characterized. The conditions of non-canonical provability thereby play no role in the meaning specification.

---

[7] However, Dummett's himself remarks that one could invert the picture by taking eliminations as meaning-giving. In a sense, inversion may be viewed as simply stating a requirement of 'harmony' between two aspects of meaning without ascribing priority to any of the two.

[8] Cf. Dummett (1991, pp. 252–256) proof-theoretic justification of second grade.

Unfortunately, this account of understanding in terms of canonical provability faces some difficulties as soon as one considers logical constants having introduction rules of the more complex kind.

### 3.3 Implication and Hypothetical Deductions

When introduction rules discharging assumptions are considered, it is no longer possible to specify what is a canonical proof of a logically complex sentence only in terms of what is a canonical proof for its sub-sentences.

By means of implication introduction rule, a proof of an implication $A \to B$ can be obtained not only from proofs of its consequent $B$, but in general also from deductions having $B$ as conclusion and the antecedent $A$ as an undischarged assumption.

As observed, a proof is a deduction in which all assumptions are discharged. A deduction in which one or more assumptions are not discharged establishes the conclusion only hypothetically.

So, in order to specify what is a proof of certain complex sentences we need to refer to deductions establishing its sub-sentence only hypothetically.

Furthermore, we can pose no restriction on the kind of rules that may appear in a hypothetical deduction leading to an implication introduction rule (what Dummett (1991, p. 260) calls 'subordinate deductions'). For example, the most direct proof of the sentence $(A \wedge B) \to A$:

$$\frac{\dfrac{[A \wedge B]}{A} \wedge\text{-elim}}{(A \wedge B) \to A} \to\text{-intro}$$

is constituted by a hypothetical deduction having $A$ as conclusion and $A \wedge B$ as assumption which consists of a single application of an elimination rule.

As this simple example shows, when coping with introduction rules discharging assumptions, "it is impossible to demand that a subordinate deduction be capable of being framed so as to appeal only to introduction rules." (*ibid.*) As a consequence, the notion of canonical proof must be relaxed, by requiring (at least in the case of implication) only the last step of the proof to be an introduction.[9]

---

[9]Prawitz (1971; 2006) and Schroeder-Heister (2006) require only the last step to be an introduction. Dummett (1991, pp. 259–64) proposes a more stringent notion, called by Prawitz (2006, §4.1) 'hereditary canonical', which is nearer to the notion of closed normal deduction, in that redundancies are forbidden in what Dummett calls the 'main stem' of the deduction. The difference is irrelevant for the aim of this paper.

But once the notion of canonical proof is so relaxed it is questionable whether it can still be claimed that canonical proofs are shaped *step by step* according to the meaning of a sentence. Rather, they are shaped according to the meaning of the sentence only in the *last* step.

As a consequence, in the case of implication, we cannot expect a canonical proofs of it to be constituted by canonical proofs of its sub-sentences, but only by a hypothetical deduction having the consequent as conclusion and the antecedent as assumption.

## 4 Spelling out the Analogy

The need of referring to hypothetical deductions in specifying the canonical proof-conditions of implications mirrors the need of referring to the mastery of complex predicates in specifying the truth-conditions of quantified sentences.[10]

The understanding of the truth-conditions of a quantified sentence depends on the mastery of a predicate contained in its instances. Where does this mastery come from?

Either we assume, with Frege, that understanding the truth-conditions of a sentence implies mastery of all the complex predicates it contains. Or, we take mastery of predicates as an independent ingredient of sentences' understanding alongside with the understanding of their truth-conditions.

Either way, understanding a sentence implies (directly or indirectly) the mastery of all the complex predicates it contains. Hence, it looks quite implausible that understanding a sentence does not imply the recognition of its alternative analyses yield by such predicates.

In the proof-conditional case, the understanding of the canonical proof-conditions of implications depends on what may be called the 'mastery' of hypothetical deductions having the consequent as conclusion and the antecedent as assumption.

Also in this case, we can ask where does this mastery come from. The structural similarity between the truth- and proof-conditional accounts of meaning marks out the very same leeway for answering this question.

---

[10]The analogy between predicates and hypothetical deduction, as well as between quantifiers and implication, could be very concisely stated in terms of the so-called Curry-Howard isomorphism between deductions in natural deduction and terms of a simply-typed lambda calculus. Implication is viewed, like the quantifiers, as a variable binding operator; and hypothetical deductions are open terms, i.e. predicates.

As a matter of fact, Dummett (following Prawitz) reduces hypothetical deductions to the proofs resulting by their applications.[11] As we observed, these fail, in general, to be canonical. But Dummett's so-called 'fundamental assumption' warrants in turn that non-canonical proofs reduce to canonical ones.[12]

This two-steps reduction of hypothetical deductions to canonical proofs mirrors Frege's assumption which reduces the mastery of complex predicates to sentences' truth-conditions. That is, the reduction amounts to assuming that knowing the canonical proof-conditions of a sentence implies (i) knowing its non-canonical proof-conditions; which in turn implies (ii) the mastery of the hypothetical deductions having the sentence as conclusion.

What is then of Dummett's claim that understanding the non-canonical proof-conditions of a sentence is unnecessary for understanding it?

As in the truth-conditional case, the rejection of this assumption may lead to an alternative answer.

The mastery of hypothetical deductions, rather than being reduced to canonical proofs, could be postulated as an independent extra-ingredient of understanding. That is, understanding a sentence would be equated to understanding its canonical proof-conditions *and* mastering the hypothetical deductions in which it figures either as conclusion or as undischarged assumption.

In the truth-conditional case, I observed that the extra-ingredient to sentences' understanding—the mastery of complex predicates—can actually supersede the other—the understanding of the truth-conditions—since the latter, Tarski *docet*, can be defined in terms of the former.

Investigating whether something analogous happens in the proof-conditional case, namely whether it is possible to recover the notion of proof from the one of hypothetical deduction, will be addressed in subsequent work.

---

[11] Cf., for instance, Prawitz (1985, p. 162): "Viewing an open argument as an argument-schema, the question of its validity should be reduced to that of a closed argument: an open argument is to be valid just in case all its closed instances obtained by substituting valid closed arguments for its free assumptions are valid."

[12] In Dummett's words, "if we have a valid argument for a complex sentence, we can construct a valid argument for it which finishes with an application of one of the introduction rules governing its principal operator." (Dummett, 1991, p. 254)

## References

Brennan, A. (2011). Necessary and sufficient conditions. In E. N. Zalta (Ed.), *The Stanford encyclopedia of philosophy* (Winter 2011 ed.).

Dummett, M. (1973). The justification of deduction. *Proceedings of the British Academy, 59*, 201–232. (Reprinted in *Truth and Other Enigmas* (pp. 290–318), 1978, London: Duckworth.)

Dummett, M. (1975). The philosophical basis of intuitionistic logic. In H. Rose & J. Shepherdson (Eds.), *Logic colloquium '73, proceedings of the logic colloquium* (pp. 5–40). Elsevier.

Dummett, M. (1981). *Frege. Philosophy of language* (2nd ed.). London: Duckworth.

Dummett, M. (1991). *The logical basis of metaphysics*. London: Duckworth.

Prawitz, D. (1971). Ideas and results in proof theory. In J. Fenstad (Ed.), *Proceedings of the second scandinavian logic symposium* (pp. 235–307). Elsevier.

Prawitz, D. (1985). Remarks on some approaches to the concept of logical consequence. *Synthese, 62*, 153–171.

Prawitz, D. (2006). Meaning approached via proofs. *Synthese, 148*, 507–524.

Schroeder-Heister, P. (2006). Validity concepts in proof-theoretic semantics. *Synthese, 148*, 525–571.

Tranchini, L. (2012). Truth from a proof-theoretic perspective. *Topoi, 31*(1), 47–57.

Luca Tranchini
Eberhard Karls Universität Tübingen
Wilhelm-Schickard Institute
Sand 13, 72076 Tübingen, Germany
e-mail: luca.tranchini@gmail.com
URL: http://sites.google.com/site/lucatranchini

www.ingramcontent.com/pod-product-compliance
Lightning Source LLC
Chambersburg PA
CBHW070735160426
43192CB00009B/1447